WILLIAM THOMSON KELVIN

NOTES OF LECTURES

ON

MOLECULAR DYNAMICS

AND

THE WAVE THEORY OF LIGHT

DELIVERED AT THE
JOHNS HOPKINS UNIVERSITY, BALTIMORE

Stenographically Reported by
A.S. Hathaway

Elibron Classics
www.elibron.com

Elibron Classics series.

ISBN 1-4021-8682-7 (paperback)
ISBN 1-4021-0939-3 (hardcover)

This Elibron Classics Replica Edition is an unabridged facsimile of the edition published in 1884 by Johns Hopkins University, Baltimore.

NOTES OF LECTURES

ON

Molecular Dynamics

and

The WAVE THEORY OF LIGHT

Delivered at the Johns Hopkins University Baltimore.

BY

SIR WILLIAM THOMSON,

Professor in the University of Glasgow

STENOGRAPHICALLY REPORTED BY

A. S. HATHAWAY,

Lately Fellow in Mathematics of the Johns Hopkins University.

1884.

BALTIMORE, MD.

ADVERTISEMENT.

In the month of October, 1884, Sir William Thomson of Glasgow, at the request of the Trustees of the Johns Hopkins University in Baltimore, delivered a course of twenty lectures before a company of physicists, many of whom were teachers of this subject in other institutions. As the lectures were not written out in advance and as there was no immediate prospect that they would be published in the ordinary form of a book, arrangements were made with the concurrence of the lecturer, for taking down what he said by short-hand.

Sir William Thomson returned to Glasgow as soon as these lectures were concluded and has since sent from time to time, additional notes which have been added to those which were taken when he spoke. It is to be regretted that under these circumstances he has had no opportunity to revise the reports. In fact he will see for the first time simultaneously with the public this repetition of thoughts and opinions which were freely expressed in familiar conference with his class. The "papyrograph" process which for the sake of economy has been employed in the reproduction of the lectures does not readily admit of corrections and some obvious slips, such as Canchy for Cauchy, have been allowed to pass without emendation; but the stenographer has given particular attention to mathematical formulas, and he believes that the work now submitted to the public may be accepted, on the whole, as an accurate report of what the lecturer said.

Lecture I.

The most important branch of physics which at present makes demands upon molecular dynamics seems to me to be the wave theory of light. When I say this, I do not forget the one great branch of physics which at present is reduced to molecular dynamics, the kinetic theory of gases. In saying that the wave theory of light seems to be that branch of physics which is most in want, which most inevitably demands applications of molecular dynamics just now, I mean that as the kinetic theory of gases is a part of molecular dynamics, is founded upon molecular dynamics works wholly within molecular dynamics, to it molecular dynamics is everything, and it must be advanced by molecular dynamics, so the wave theory of light is only beginning to demand imperatively applications of that kind of dynamical science.

The wave theory of light began very much in the hands of Fresnel, afterwards of Cauchy, and to some degree though not perhaps to so great a degree, in the hands of Green. It was wholely molecular dynamics, but of an imperfect kind in the hands of Fresnel. Cauchy attempted to found his mathematical investigations on a molecular treatment of the subject. Green almost wholly shook off the molecular treatment, and worked out all that was to be worked out in that way for the wave theory of light, by the dynamics of continuous matter. Indeed, I do not know that it is possible to add substantially to what Green has done in this subject. Substantial additions are scarcely to be made to a thing that is applied as Green's work is, on the explanation of the propagation of light, the refraction and the reflection of light at the bounding surface of two different mediums, and the propagation of light through crystals, by a strict mathematical treatment, founded on the consideration of homogeneous, elastic matter. Green's treatment is really complete in this respect, and there is nothing substantial to be added to it. But there is a great deal of exposition wanting to let us make it our own. We must study it; we must try to see what there is in the very concise and sharp treatment, with some very long formulae, which we find in Green's papers.

The wave theory of light, treated on the assumption that the medium through which the light is propagated is continuous and homogeneous, except where distinctly separated by a bounding inter-face between two different mediums, is really completed by Green. But there is a great deal to be learned from that kind of treatment that perhaps scarcely has yet been learned, because the subject has not been much studied and reduced to a very popular form hitherto.

Cauchy seemed unable to help beginning with the consideration of discreet particles mutually acting upon one another. But, except in his theory of dispersion he virtually came to the same thing somewhat soon in his treatment everytime he began it afresh, as if he had commenced right away with the consideration of a homogeneous, elastic solid. Green preceded him, I believe, in this subject. I read a statement of Lord Rayleigh that there seems to have been a matter of fact attributing to Cauchy of that which Green had actually done before. Green had exhausted the subject. but there is no doubt that Cauchy worked in an independent way.

What I propose in this first Lecture — we must have a little mathematics, and I must not be too long with any kind of preliminary remarks — is to call your attention to the outstanding difficulties. The first difficulty that meets us in the dynamics of light is the explanation of dispersion, that is to say, of the fact that the velocity of propagation of light is different for different wave lengths or for light of different periods in one and the same medium. Treat it as we will, vary the fundamental suppositions as much as we can, as much as the very fundamental idea allows us to vary them, and we cannot force from the dynamics of a homogeneous elastic solid a difference of velocity of wave propagation for different periods.

Cauchy pointed out that if the spheres of action of individual molecules be comparable with the wave lengths the fact of the difference of velocities for different periods or for different wave lengths in the same medium is explained. The best way, perhaps, of putting Cauchy's fundamental explanation is to say that there is hetreogeneousness through space, comparable with the wave length in the medium — that is, if we are to explain dispersion by Cauchy's unmodified supposition. We shall consider that a little later. I have no doubt it is perfectly familiar already to many of you that it is essentially insufficient to explain the facts.

Another idea for explaining dispersion has come forward more recently, and that is the assumption of molecules loading the luminiferous ether and somehow or other elastically connected with it. The first distinct statement that I have seen of this view is in Helmholtz's little paper on anomalous dispersion. I shall have occasion to speak of that a good deal and to mention other names whom Helmholtz quotes in this respect, so that I shall say nothing about it historically, except that there we have in Helmholtz's paper and by some German mathematicians who preceded him quite another departure in respect to the explanation of dispersion. The Cauchy hypothesis gives us something comparable with the wave length in the geometrical dimensions of the body. Or, to take a crude matter of fact view of it, let us say the ratio of the distance from molecule to molecule (from the center of one molecule to the center of the next nearest molecule) to the wave length of light is the fundamental characteristic, as it were, to which we must look for the explanation of dispersion upon Cauchy's theory.

We may take this fundamental idea in connection with the two hypothesis for accounting for dispersion: that we must have some relation either to wave length or to period, and it seems (altho' this is a proposition that would require modification) at first sight that with very long waves the velocity of propagation should be independent of the period or wave length. That, at all events seems to be the case when the subject is only looked upon according to Cauchy's view. We are led to say then that it seems that for very long waves there should be a constant velocity of propagation. Experiment and observation now seems to be falling in very distinctly to affirm the conclusions that follow from the second hypothesis that I alluded to to account for dispersion. In this second hypothesis, instead of having a geometrical dimension in the solid which is comparable with the wave length, we have a fundamental time relation — a certain definite interval of time somehow ingrained in the constitution of the solid with a definite relation to the period. So that, instead of a relation of length to length, we have a relation of time to time.

Now, how are we to get our time element ingrained in the constitution of matter? We can scarcely put that question now-a-days. We are all familiar with the time of vibration of the sodium atom, and the great wonders revealed by the spectroscope are all full of indications showing a relation to absolute intervals of time in the properties of matter. This is now as well understood, that it is no new idea to propose to adopt as our unit of time one of the fundamental periods — for instance, the period of vibration of light in one or other of the sodium **D** lines. You all have a dynamical idea of this already. You all know something about the time of vibration of a molecule, and how the time of vibrabration of light in passing through any substance in supposing it nearly the same as the natural time of vibration of the molecules of the substance, gives rise to the absorption. We all know of course, according to this idea, the old dynamical explanation, ~~first~~ proposed by Stokes, of the dark lines of the solar spectrum.

We have now this interesting point to consider ~~that~~ if we would work out the idea of dispersion at all, we must look definitely to times of vibration, in connection with solid itself. To get a firm hypothesis that will allow us to work at the subject, let us imagine the luminiferous ether occupied by something different from the luminiferous ether itself. That something might be a portion of denser ether, or a portion of more rigid ether: or we might suppose a portion of ether to have greater density and greater rigidity, or different density and different rigidity from the surrounding ether. We will come back to that subject in connection with the explanation of the blue sky, and, particularly, Lord Rayleigh's dynamics of the blue sky. In the meantime, I want to give something that will allow us to bring out a very crude mechanical model of dispersion.

In the first place, we must not listen to any suggestion that we must look upon the luminiferous ether as an ideal way of putting the thing. A real matter between us and the remotest stars I believe there is, and that light consists of real motions of that matter motions just such as are described by Fresnel and Young, motions in the way of transverse vibrations. If I knew what the magnetic theory of light is, I might be able to think of it in relation to the fundamental principles of the wave theory of light. But it seems to me that it is rather a backward step from an absolutely definite mechanical motion that is put before us by Fresnel and his followers to take up the so-called Electo-magnetic theory of light in the way it has been taken up by several writers of late. In passing I may say that the one thing about it that seems intelligable to me, I scarcely think is admissable. What I mean is, that there should be an electric displacement perpendicular to the line of propagation and a magnetic disturbance perpendicular to both. It seems to me that when we have an electro-magnetic theory of light, we shall see electric displacement as in the direction of propagation — simple vibrations as described by Fresnel with lines of vibration perpendicular to the line of propagation — for the motion actually constituting light. I merely say that in passing, as perhaps some apology is necessary for my insisting upon the plain matter of fact dynamics and the true elastic solid as giving what seems to me the only tenable foundation for the wave theory of light in the present state of our knowledge.

The luminiferous ether we must imagine to be a substance which so far as luminiferous vibrations are concerned moves as if it were an elastic solid. I do not say it is an elastic solid. That it moves as if it were an elastic solid in respect to the luminiferous vibrations is the fundamental assumption of the wave theory of light.

An initial difficulty that might be considered inseparable is, how can we have an elastic solid, with a certain degree of rigidity pervading all space, and the earth moving through it at the rate the earth moves around the sun, and the sun and solar system moving through it at the rate in which they move through space, at all events relatively to the other stars.

That difficulty does not seem to me so very insuperable. Suppose you take a piece of Burgundy pitch, or Trinidad pitch, or what I know best for this particular subject, Scotch shoemakers wax. That is the substance I used in the illustration I intend to refer to. I do not know how far the others would succeed in the experiment. Suppose you take one of these substances, the shoemakers wax, for instance. It is brittle, but you can bend it into the shape of a tuning fork and make it vibrate. Take a long rod of it, and you can make it vibrate as if it were a piece of glass. But leave it lying upon its side for a night and it will flatten down gradually. The weight of a letter will flatten it. Experiments have not been made as to the fluidity or non-fluidity of such a substance as shoemaker's wax; but that time is all that is necessary to allow it to yield absolutely as a fluid is not an improbable supposition with reference to any one of the substances I have mentioned. Scottish shoemaker's wax I have used in this way: I took a large slab of it, perhaps a couple of inches thick, fitting in a glass jar ten or twelve inches in diameter. I filled the glass jar with water and laid the slab of wax in it with a quantity of corks underneath and two or three lead bullets on the upper side. This was at the beginning of an Academic year. Six months passed away and the lead bullets had all disappeared and I suppose the corks were half way through. Before the year had passed on looking at the slab I found that the corks were floating in the water at the top, and the bullets of lead were tumbling about in the bottom of the jar.

Now, if a piece of cork, in virtue of the greater specific gravity of the shoemaker's wax would float upwards through that solid material and a piece of lead, in virtue of its greater specific gravity would move downwards through the same material, though only at the rate of an inch per six months we have an illustration, it seems to me, quite sufficient to do away with with the fundamental difficulty from the wave theory of light. Let the luminiferous ether be looked upon as a wax which is elastic and I was going to say brittle, (we will think of that yet of what the meaning of brittle would be) and capable of executing vibrations like a tuning fork when times and forces are suitable — when the times in which the forces tending to produce distortion act, are very small indeed, and the forces are not too great to produce rupture. When the forces are long continued then very small forces suffice to produce change of shape. Whether infinitessimally small forces produce change of shape or not we do not know; but very small forces suffice to produce change of shape. All we have got with respect to the luminiferous ether is that the exceedingly small forces required to be brought with play in the luminiferous vibrations do not in the times during which they act suffice to produce any sensibly permanent distortion. The come and go effects taking place in the period of the luminiferous vibrations do not give rise to the consumption of any large amount of energy, not large enough an amount to cause the light to be wholly absorbed in say its propagation from the remotest visible star to the earth.

If we have time, we shall try a little later to think of some of the magnitudes concerned, and think of, in the first place, the magnitude of the shearing force in luminiferous vibrations of some assumed amplitude, on the one hand, and the magnitude of the shearing force concerned, when the earth, say moves through the luminiferous ether on the other hand. The subject has not been gone into very fully; so that we do not know at this moment whether the earth moves dragging the luminiferous ether altogether with it, or whether it moves more nearly as if it were through a frictionless fluid. It is conceivable that it is not impossible that the earth moves through the luminiferous ether almost as if it were moving through a frictionless fluid and yet that the luminiferous ether has the rigidity necessary for the performance of the luminiferous vibration is the period from the four hundred million millionth of a second to the eight hundred million millionth of a second corresponding to the visible rays, or from the periods which we now know in the low rays of radiant heat as recently experimented on and measured for the wave length by Abney, to the high ultra violet rays of light known chiefly by their chemical actions. If we consider the exceeding smallness of the period from the 100 million millionth of a second to the 1600 million millionth of a second through the known range of radiant heat and light, we need not fully despair of understanding the property of the luminiferous ether. It is no greater mystery at all events than the shoemaker's wax. That is a mystery, as all matter is; the luminiferous ether is no greater mystery.

We know the luminiferous ether better than we know any other kind of matter in some particulars. We know it for its elasticity. we know it in respect to the constancy of the velocity of propagation of light for different periods. Take the eclipses of Jupiter's satellites or something far more telling yet, the bursting of luminous stars and so on, as referred to by Prof. Newcomb in a recent discussion at Montreal on the subject of the velocity of propagation of light in the luminiferous ether. These phenomena prove to us with tremenduously searching test, to an excessively minute degree of accuracy, the constancy of the velocity of propagation of all the rays of visible light through the luminiferous ether.

Luminiferous ether must be a body of most extreme simplicity. It may be perhaps soft. We might imagine it to be a body whose ultimate property is to be incompressible; to have a definite rigidity for vibrations in times less than a certain limit, and yet to have the absolutely yielding character that we recognize in wax-like bodies when the force is continued for a sufficient time

It seems to me that we must know a great deal more of the luminiferous ether than we do. But instead of beginning with saying that we know nothing about it I say that we know more about it than we do about air or water, glass or iron — it is far simpler, there is far less to know. That is to say, the natural history of the luminiferous ether is an infinitely simpler subject than the natural history of any other body. It seems probable that the molecular theory of matter may be so far advanced sometime or other that we can understand an excessively fine-grained structure and understand the luminiferous ether as differing from glass and water and metals in being very much more finely grained in its structure. We must not attempt, however, to jump too far in the inquiry, but take it as it is, and take the great facts of the wave theory of light as giving us strong foundations for our convictions as to the luminiferous ether.

Imagine for a moment that we make a rude mechanical model. Let this be an infinitely rigid spherical shell; let there be another absolutely rigid shell inside of that, and so on, as many as you please. Naturally, we might think of something more continuous than that, but I only wish to call your attention to a crude mechanical explanation, possibly of the effects of dispersion. Suppose we had luminiferous ether outside, and that this hollow space

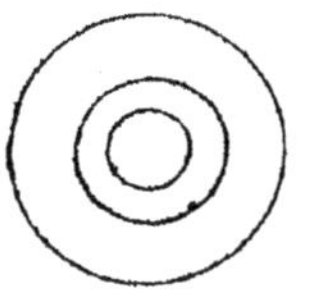

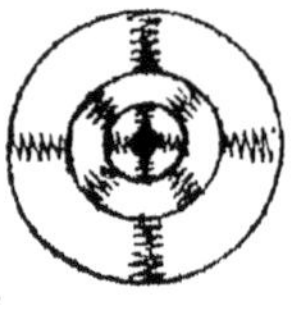

is of very small diameter in comparison with the wave length. Let zig-zag springs connect the outer rigid boundary with boundary number two. I use a zig-zag, not a spiral spring, which observes the helical properties which we are not ready for yet, such properties as sugar and quartz have in disturbing the luminiferous vibrations. Suppose we have shells 2 and 3 also connected by a sufficient number of zig-zag springs and so on; and let there be a solid enclosed in the center with spring connections between it and the shell outside of it. If there is only one of these interior shells, you will have one definite period of vibration. Suppose you take away everything except that one interior shell; displace that shell and let it vibrate. The period of its vibration is perfectly definite. If you have an immense number of such shells, with movable molecules inside of them distributed through some portion of the luminiferous ether, you will put it into a condition in which the velocity of the propagation of the wave will be different from what it is in the homogeneous luminiferous ether. You have what is called for, viz., a definite period; and the relation between the period of vibration in the light considered and the period of the free vibration of the shell will be fundamental in respect to the attempt of a mechanism of that kind to represent the phenomena of dispersion.

If you take away everything except the one shell, you will have almost exactly, I think, the view of Helmholtz's paper— a crude model, as it were, of what Helmholtz makes his paper on anomalous dispersion. Helmholtz, besides that, supposes a certain degree or coefficient of viscous resistance against the vibration of the inner shell, relatively to the outer one. Helmholtz does not reduce it to a gross mechanical form like this, but merely assumes particles connected with the luminiferous ether and assumes a viscous motion to operate against the motion of the particles.

If we had only absorption and dispersion to deal with, there would be no difficulty whatever in accounting for all that is necessary. When the period of luminiferous vibration is smaller than the natural vibration of the first shell, we have a certain state of things; when it is the same we have what is prettiest, the mathematical conditions of absorption and the infinite vibrations are wanting. What is meant by absorption in the interior? The conversion of luminiferous vibrations into heat or some other mode of action or the dynamics must be such that when the motion of vibration of this inner shell is through a greater and greater range the period ceases to fulfill the conditions of exactness, and so, without absorption the infinity vibration is not met with. This part of the subject will occupy us more fully a little later.

If we had only dispersion to deal with there would be no difficulty in getting a full explanation by putting this not in a rude mechanical model form, but in a form which would commend itself to our judgment as presenting the actual mode of action of the particles, whatever they may be upon the particles of luminiferous ether. We except the heavier matter; but oxygen, hydrogen and such as those must somehow or other act in the luminiferous ether, have some sort of elastic connection with it. and I cannot imagine anything that commends itself to our ideas better than this sort of thing. By taking enough of these interior shells, and by neglecting the idea of absolute continuity with no limit whatever to the period we may come as it were to the kind of mutual action that exists between any particular atom and the luminiferous ether. It seems to me that there must be something in this, that this, as a symbol, is certainly not an hypothesis, but a certainty.

But alas for the difficulties of the undulatory theory of light, refraction and reflection at plane surfaces worked out by Green differ in the most irreducable way from the facts. They correspond in some degree to the facts, but there are differences that we have no way of explaining at all. A great many hypothesis have been presented, but none of them seems at all tenable.

First of all is the question, are the vibrations of light perpendicular to, or are they in the plane of polarization - defining the plane of polarization as the plane through the incident and refracted rays, for light polarized by reflection. Think of light polarized by reflection at a plane surface and the question is, are the vibrations in the reflected ray perpendicular to the plane of incidence and reflection, or are they in the plane of incidence and reflection. I merely speak of this subject in the way of index. We shall consider very fully, Green's theory and Lord Rayleigh's work upon it. I come to the conclusion with absolute certainty, it seems to me, that the vibrations must be perpendicular to the plane of incidence and reflection of the light that is polarized by reflection.

Now there is this difficulty outstanding – the theory which gives this result does not give it rigorously, but only approximately. We have by no means so good an approach in the theory to complete extinction of the vibrations in the reflected ray (when we have the light in the incident ray vibrating in the plane of incidence and reflection) as observation gives. I shall say no more about that difficulty, because it will occupy us a good deal later on, except to say that the theoretical explanation of reflection and refraction is not satisfactory. It is not complete, and it is unsatisfactory in this, that we do not see any way of mending it.

But suppose for a moment that it might be mended and there is a question connected with it which is this: Is the difference between two mediums a difference corresponding to difference of rigidity, or does it correspond to difference of density. That is an interesting question, and some of the work that was done upon it seemed most tempting in respect to the supposition that the difference between two mediums is a difference of rigidity and not a difference of density. When fully examined, however, the seemingly plausible way of explaining the facts of refraction and reflection by difference of rigidity and no difference of density is found to be delusive, and we are forced to the view that there is difference of density and very little difference of rigidity.

It seemed to me in working out this subject very carefully, and endeavoring to understand Lord Rayleigh's work upon it, and learn what had been done by others, for a time to be too much of an assumption that the rigidity was exactly the same and that the whole effect was due to difference of density. Might it not be, it seemed to me, that the luminiferous ether on the two sides of interface at which the refraction and reflection takes place, might differ both in rigidity and in density. It seemed to me then by a piece of work (which I must verify, however, before I state quite confidently about it) that by supposing the luminiferous ether in the commonly called denser medium to be considerably denser than it would be where the rigidity is equal, and the rigidity to be greater than in the other mediums, that we might get a better explanation of the polarization by reflection than Green's result gives. Green's work ends with the supposition of equal rigidities and unequal densities. He puts the whole in his formulae to begin with, but he ends with this supposition and his result depends upon it.

Not to deal in generalities, let us take the case of glass and a vacuum, say. It seemed to me that by supposing the rigidity of luminiferous ether in glass to be greater than in vacuum and the density to be greater, but greater in a greater proportion than the rigidity, so that the velocity of propagation is less in glass than in vacuum, that we should get a better explanation of the details of polarization by reflection than Greens result gives.

It is only since I have left the other side of the Atlantic that I have worked at this thing, and going at it with considerable interest, I enquired of everybody I met whether there were any observations that would help me. At last I was told that Prof. Rood had done what I desired to know, and on looking at his paper, I found that it settled the matter.

My question was this: Has there been any measurement of the intensity of light reflected at nearly normal incidence from glass or water considerably greater than Fresnel's formula gives. Fresnel gives $\left(\frac{\mu-1}{\mu+1}\right)^2$ for the ratio of the intensity of the reflected ray to the intensity of the incident ray in the case of normal incidence, or incidence nearly normal. I wanted to find out whether that had been verified. It seems that nobody had done it at all until Prof. Rood, of Columbia College, New York, took it up. His experiments showed to a rather minute degree of accuracy an agreement with Fresnel's formulae, so that the explanation I was inclined to make was disproved by it. I myself had worked with the reflection of a candle from a window glass and had come to the same conclusion, through such very crude and rough approximate results. At all events, I satisfied myself that there was not so great a deviation from Fresnel's law as would allow me to explain the difficulties of refraction and reflection by assuming greater rigidity, for example, in glass than in air. We are now forced very much to the conclusion from several results, but directly from Prof. Rood's photometrical experiments, that the rigidity must be very nearly equal in the two.

There is quite another supposition that might be made that would give us the same law, the supposition that the reflection depends wholly upon difference of rigidity and that the densities are equal in the two. That gives rise to the same intensity of reflected light, so that the photometric measurement does not discriminate between the two extremes, but it does prevent us from pushing in on the other side of generally accepted result in the manner that I had thought of.

We may look upon the explanation of polarization, by reflection and refraction as not altogether unsatisfactory, although not quite satisfactory, and you may see that this kind of modification of the luminiferous ether is just what would give us the virtually greater density. How this gives us precisely the same effect as a greater density I shall show when we work the thing out mathematically. We shall see that this supposition is equivalent to giving the luminiferous ether a greater density, without making the addition to the density according to the idea of vibration.

I am approaching an end I had hoped to get sooner. We have the subject of double refraction in crystals, and here is the great hopeless difficulty. I do not find it quite correctly stated even in places in which it is referred to. For instance, even Lord Rayleigh says, that Fresnel's view requires us to suppose the rigidity of the luminiferous ether to depend on the direction of the vibration — which is not quite true. The rigidity cannot depend on the direction of the vibration.

If we look into the matter of the distortion of the elastic solid, we may consider, possibly, that that is not wonderful; but Fresnel's supposition as to the direction of the vibration of light is that the conclusion that the plane of vibration is perpendicular to the plane of polarization proves, if it is true, that the velocity of propagation of light in uniaxial crystals depends on the direction of vibration and not on the plane of the distortion. In the vibrations of light, we have to consider the medium as being distorted and tending to recover its shape.

Let this be a piece of uniaxial crystal iceland spar, for instance, a round or square column, with its length in the direction of the optic axis, which I will represent in the board by a dotted line.

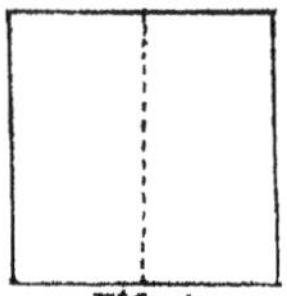

FIG. 1

Now the relation between light polarized by passing through iceland spar on the one hand and light polarized by reflection on the other hand shows us that if the line of vibration is perpendular to the plane of polarization, then the velocity of propagation of light in different directions through iceland spar depends solely on the line of vibration and not at all on the plane of distortion.

There is no way in which that can be explained by the rigidity of an elastic solid. Look upon it in this way, in the first place. Take a cube of iceland spar, keeping the same direction of the axis as before. Let the light be passing downwards, as indicated by the dotted arrowheads. What would be the mode of vibration, with such a direction of propagation? Let us suppose, in the first place, the

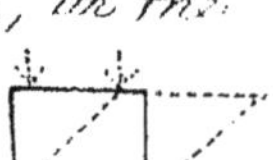

Fig 2.

vibrations to be in the plane of the diagram. Then the distortion of that portion of matter will be in the direction indicated; a portion which was rectangular swings into shape represented by the dotted lines. The force tending to cause a piece of matter which has been displaced to resume its original shape depends on this kind of distortion. The mathematical expression of it would be n a constant of rigidity, multiplied into a, the amount of the distortion. How that is to be reckoned is familiar to many of you, and we will not enter into the details just now. But just consider this other case, where the direction of propagation of the light is horizontal, as indicated by the dotted arrow-heads, that is to say, propagated perpendicular to the axis of the crystal (Fig. 3). What would be the nature of the distortion here the vibration being still on the plane of the diagram? The distortion will be in this way in which I move my two hands. A portion which was rectangular will swing into this shape indicated by the dotted lines. The return force will then depend upon a distortion of that kind. But a distortion of that kind (Fig. 3), is identical with a distortion of this kind (Fig. 2) and the result must be, if the effect depends upon the return force in an elastic solid, that we must have the same velocity of propagation in this case and in this case (Figs 2 and 3).

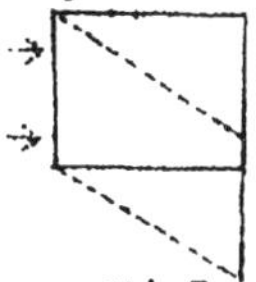

Fig. 3.

But observe this is the case of the extraordinary ray; and you know that we have greater velocity of propagation in the first case, and less in the second. There is an outstanding difficulty that is absolutely inexplicable on the bare theory of an elastic solid.

The question now occurs, may we not explain it by loading the elastic solid. But the difficulty is, to load it unequally in different directions. Lord Rayleigh thought that he had got an explanation of it in his paper to which I have referred. He was not aware that Rankine had exactly the some idea. Lord Rayleigh at the end of this paper puts forward the supposition that difference of effective inertias in different directions may be adduced to explain the difference of velocity of propagation in iceland spar. But if that were the case the wave propagation would not follow Huygen's law. It would follow the law according to which the velocity of propagation would be inversely proportional to what it is according to Huygen's law. Huygen's geometrical construction for the extraordinary ray in iceland spar gives us an ellipsoid of revolution according to which the velocity of propagation of light will be found by drawing from the centre of the ellipsoid a perpendicular to the tangent plane. For example (Fig. 4) CN will correspond to the velocity of propagation of the light when the front is in the direction of this (tangent) line. If the velocity is different in different directions in virtue of an effective inertia, as your Instructor will humbly hold, Lord Rayleigh's idea is that the vibrating molecules might be like oblate spheroids vibrating in a frictionless fluid. It will have greater effective inertia when vibrating in the direction of its axis perpendicular to its flat side, less effective inertia when vibrating in its equatorial plane. That is a very beautiful idea, and we absolutely want it to explain the difficulty if the pushing forward of the conclusions from it were verified by experiment. Stokes has made the experiment. He did not know of Rankine's paper. Rankine made the first suggestion in the matter but did not push the question further than to give it as a mode of getting over the difficulty in double refraction. Stokes took away the poetry of it. He experimented on the refracting index of iceland spar for a variety of incidences, and found with minute accuracy indeed that Huygens construction was verified and that therefore it was impossible to account for the unequal velocity of propagation in different directions by the beautiful suggestions of Rankine and Lord Rayleigh.

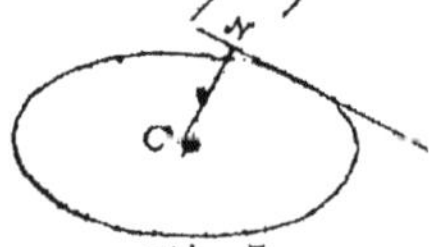

Fig 4.

I have not been able to make a suggestion, but I have great hopes that these spring arrangements are going to work us out of the difficulty. I will, just in conclusion, give you the idea of how it might.

We can easily suppose these spring arrangements to have different strengths in different directions; and their law will suit exactly. Their law will give the fundamental thing we want which is that the velocity of propagation of light shall depend on the direction of vibration, and not on the distortion. Besides that, this will obviously verify Huygens's law – it gives us exactly the same law as the elastic theory gives.

But alas, alas, we have one difficulty which seems still insuperable and prevents my putting this forward as the explanation, and that is, that I cannot get the requisite difference of effective inertia in different directions for the different wave lengths to suit. If we take this theory, we should have, instead of the very nearly equal difference of refractive index for the different rays in such a body as iceland spar, with dispersion merely a small thing in comparison with those differences, that the difference of refractive index in different directions would be comparable with dispersion and modified by dispersion to a prodigious degree, and in fact we should have anomalous dispersion coming in between the velocity of propagation in one direction and the velocity of propagation in another. The impossibility of getting a sufficiently constant difference of wave velocity in different directions for the different periods in those directions seems to me to be a puzzler.

So now, I have given you one hour and seven minutes and brought you face to face with a difficulty which I will not say is insuperable, but something in which nothing ever has been done from the beginning of the world to the present time that will give us the slightest explanation.

I shall do to-morrow, what I had hoped to do to-day, give you a little mathematics, knowing that it is not going to explain everything, but I think we will have an interest in working out the motions of an elastic solid and obtaining a few solutions that depend on the equations of motion of an elastic solid. I shall first take the case of zero rigidity; that will give us sound. We shall take the most elementary sounds possible, namely a spherical body alternately expanding and contracting. We shall pass from that to the case of a single globe vibrating to and fro in air. We shall pass from that to the case of a tuning fork, and endeavor to explain the zones of silence which you all know in the neighborhood of a vibrating tuning fork. I hope we shall be able to get through that in a short time and pass on our way to the corresponding solutions of the motions of a wave proceeding from a center in respect to the wave theory of light.

Lecture II.

In the first place, I will take up the equations of motion of an elastic solid. I assume that the fundamental principles are familiar. At the same time, I should be very glad if any person present would, without the slightest hesitation ask for explanations, if anything is not understood. I want to be at once on a Professorial footing with you, so that the work shall be rather something between you and me than something in which I shall be making a performance before you in a matter in which many of you may be quite as competent as I am, if not more so.

I want if we can get something done in half an hour, on these problems of molar dynamics as we may call it, to distinguish from Molecular dynamics, to come among you for a few moments and then go on to a problem of molecular dynamics to prepare the way for motions of mutual interference among particles under varying circumstances that may perhaps have applications in physical science and particularly to the theory of light.

The fundamental equations of equilibrium of elastic solids are, of course, included in D'Alembert's form of the equations of motion. I shall keep to the notation that is employed in Thomson and Tait's Natural Philosophy, which is substantially the same notation as is employed by other writers.

Let a, b, c, denote distortion, viz:– a, is a distortion in the plane perpendicular to OX produced by slippings of the two planes which intersect in OX.

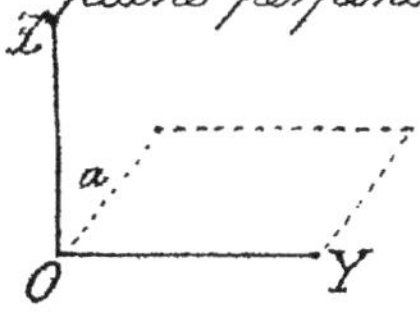

Let us consider this state of strain in which, without other change, a portion of the solid in the plane yz which was a square section becomes a rhombic figure. The measurement of that state of strain is given very fully in Thomson and Tait's geometrical preliminary for the theory of elastic solids. It is called a simple shear. It may be measured either by the rate of shifting of parallel planes per unit distance perpendicular to them, or, which comes to exactly the same thing, the change of the angle measured in radians. Then I shall put down inside this small angular space the letter a, to denote the angle measured in radians.

I use the word "radians"; it is not a very common word; I suppose you know what I mean. In Cambridge in the olden time we used to have a very illogical nomenclature, viz: "the unit angle"— a very absurd use of the article "the". It is illogical to talk of an angle being measured in "the" unit angle; there is no such thing as measuring anything, except in terms of "a" unit. The unit in which it is convenient to measure angles in Analytical Mechanics is the angle whose arc is radius. That used to be called the unit angle. My brother James Thomson proposed to call it the radian.

There are three principal distortions, a, b, c, relative to the axes of OX, OY, OZ; and again, three principal dilatations — condensations of course if any one is negative — e, f, g, which are the ratios of the augmentation of length to the length.

The general equation of energy will of course be an equation in which we have a quadratic function of e, f, g, a, b, c, the expression for which will be $\frac{1}{2}(11e^2 + 12ef + 13eg + 14ea + 15eb + 16ec + 21ef + 22f^2 + 23fg + \ldots\ldots)$

We do not deal with 11, 12, .. etc., as numbers but as representing the twenty-one coefficients of this quadratic subject to the conditions $12 = 21$, etc. If we denote this quadratic function by E, then

$$\frac{dE}{de} = 11e + 12f + 13g + 14a + 15b + 16c.$$

This is a component of the normal force required to produce this compound strain e, f, g, a, b, c. According to the notation of Thomson & Tait, let

$P = \frac{dE}{de}$, $Q = \frac{dE}{df}$, $R = \frac{dE}{dg}$, $S = \frac{dE}{da}$, $T = \frac{dE}{db}$, $U = \frac{dE}{dc}$. We have, then, the relation $Pe + Qf + Rg + Sa + Tb + Uc = 2E$ the well known dynamical interpretation of which you are of course familiar with. A little later we shall consider these 21 coefficients, first, in respect to the relations among them which must be imposed to produce a certain kind of symmetry relative to the three rectangular axes; and then see what further conditions must be imposed to fit the elastic solid for performing the functions of the luminiferous ether in a crystal.

Before going on to that we shall take the case of a perfectly isotropic material. We can perhaps best put it down in tabular form in this way.

	1	2	3	4	5	6
1	A	B	B	0	0	0
2	B	A	B	0	0	0
3	B	B	A	0	0	0
4	0	0	0	n	0	0
5	0	0	0	0	n	0
6	0	0	0	0	0	n

In the first place in this square which has to do with the distortions a, b, c, alone, if we let n represent the rigidity modulus the three diagonal terms will each be n, and those outside the diagonal will be zero. Six of the coefficients are thus determined. With reference to the upper right, and lower left, hand corner squares, let us consider what possible relations there can be for an isotropic body between longitudinal strains and distortions. Clearly none. No one of the longitudinal strains can call into play a tangential force in any of the faces; and conversely, if the medium be isotropic no distortion produced by slipping in the faces parallel to the principal planes can introduce a longitudinal stress – a stress parallel to any of the lines OX, OY, OZ. Therefore we have zeros in those squares. We know that 11 = 22 = 33 and each of these will be represented by Saxon A ($\mathfrak{A}$). Now consider the effect of a longitudinal pull in the direction of OX. If the body be only allowed to yield longitudinally, that clearly will give rise to a negative pull in the directions parallel to Oy, Oz. We have then a cross connection between pulls in the directions OX, OY, OZ. Isotrophy requires that the mutual relations must be all equal, so that we have just one coefficient to express these relations. That coefficient is denoted by Saxon B. ($\mathfrak{B}$); and that fills up our 36 squares, which represent but 21 coefficients in virtue of the relations 12 = 21, etc. We can now write down our quadratic expression for the energy,

$$E = \tfrac{1}{2}\left[\mathfrak{A}\left(e^2+f^2+g^2\right) + 2\,\mathfrak{B}\left(fg+ge+ef\right) + n\left(a^2+b^2+c^2\right)\right].$$

Instead of these Saxon letters $\mathfrak{A}, \mathfrak{B}$, which have very distinct and obvious interpretations, we may introduce the resistence of the solid to compression, the reciprocal of what is commonly called the compressibility – or perhaps to avoid ambiguities, what we may call the bulk modulus, k. Then it is proved in Thomson & Tait, and in an article in the Encyclopaedia Britannica which perhaps some of you may have that $\mathfrak{A} = k + \frac{4}{3}n$, $\mathfrak{B} = k - \frac{2}{3}n$. The considerations which show the above relations with the bulk modulus also show us that we must have $n = \frac{1}{2}(\mathfrak{A} - \mathfrak{B})$. This is most important. Take a solid cube with its three axes parallel to OX, OY, OZ. Apply a pull along two faces perpendicular to OX and an equal pressure on two faces perpendicular to OY; that will give a distortion in the plane xy. Find the value of that simple shear; it is done in a moment. Find the shearing force required to produce it calculated from $\mathfrak{A}$ and $\mathfrak{B}$ and equate that to the force calculated from the rigidity modulus n and then you find this relation. The relations for complete isotropy are exhibited here in this quadratic expression for the energy, with the equation $n = \frac{1}{2}(\mathfrak{A} - \mathfrak{B})$.

We shall pass on to the formation of the equations of motion. For equilibrium, the force applied at any point x, y, z of the solid, reckoned per unit of bulk at that point must be equal to $\left(\frac{dP}{dx} + \frac{dU}{dy} + \frac{dT}{dz}\right)$. If the body be held distorted in any way by bodily forces applied all through the interior the resultant of the elastic force on an infinitesimal portion of matter at the point x, y, z is obviously $\left(\frac{dP}{dx} + \frac{dU}{dy} + \frac{dT}{dz}\right)\delta x \delta y \delta z$. If the pull P augments as you go forward in the direction OX, there will, in virtue of that, be a resultant forward pull $\frac{dP}{dx}$ upon the infinitessimal element. On the other hand U is the stress corresponding to rotation around the axis OZ

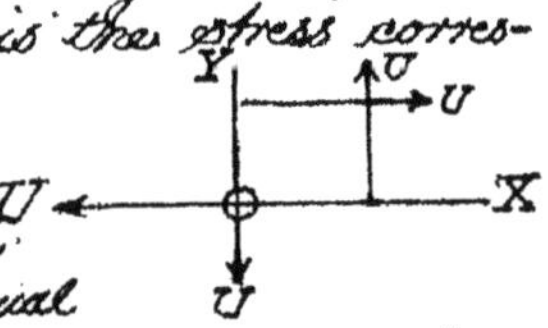

The two tangential forces, U perpendicular to OY, and U perpendicular to OX on the one pair of forces and the pair of forces is equal and in opposite directions on the other faces constitute two balancing couples, as it were. If the force parallel to OX increases as we proceed in the direction y positive, there will be a resultant positive force on this element, because it is pulled to left by the smaller and to right by the larger, and there will be an augmentation to the force in the direction of OX by $\frac{dU}{dy}$. Quite similarly, $\frac{dT}{dz}$ is a contribution to the force parallel to OX.

Now, let there be no bodily forces acting through the material, but let the inertia of the moving part, and the reaction against acceleration in virtue of inertia constitute the equilibrating reaction against elasticity. The result is, that we have the equation $\frac{dP}{dx} + \frac{dU}{dy} + \frac{dT}{dz} = \rho \frac{d^2\xi}{dt^2}$, if by ρ we denote the density and by ξ we denote the displacement from equilibrium in the direction OX of that portion of matter having x, y, z for coordinates of its mean position.

I said I would use the notation of Thomson and Tait who employ α, β, γ to denote the displacements; but errors are too common when α and a are mixed up especially in print, so we will take ξ, η, ζ, instead. I have had volumes of trouble in reading Helmholtz's paper on anomalous dispersion, on this account very frequently not being able to distinguish with a glass whether a certain letter was a or α.

The values of S, T, U, we had better write out in full, although the others may be obtained from the value of any one by symmetry. The expenditure of chalk is often a saving of brains. They are:

$$S = n\left(\frac{d\eta}{dz} + \frac{d\zeta}{dy}\right), \quad T = n\left(\frac{d\xi}{dz} + \frac{d\zeta}{dx}\right), \quad U = n\left(\frac{d\xi}{dy} + \frac{d\eta}{dx}\right).$$

We have $P = \mathfrak{A}e + \mathfrak{B}(f+g)$. There are two or three other forms which are convenient in some cases and I will put them down (writing $\underline{m}$ for $k + \frac{1}{3}n$) $P = (m+n)\frac{d\xi}{dx} + (m-n)\left(\frac{d\eta}{dy} + \frac{d\zeta}{dz}\right) = (m-n)\left(\frac{d\xi}{dx} + \frac{d\eta}{dy} + \frac{d\zeta}{dz}\right) + 2n\frac{d\xi}{dx} = m\left(\frac{d\xi}{dx} + \frac{d\eta}{dy} + \frac{d\zeta}{dz}\right) + n\left(\frac{d\xi}{dx} - \frac{d\eta}{dy} - \frac{d\zeta}{dz}\right)$ We shall denote very frequently by δ the expression $\frac{d\xi}{dx} + \frac{d\eta}{dy} + \frac{d\zeta}{dz}$, so that for example, the second of these expressions is $P = (m-n)\delta + 2n\frac{d\xi}{dx}$. If we want to write down the equations of a heterogeneous medium, as will sometimes be the case, especially in following Lord Rayleigh's work on the blue sky, we must keep these symbols $\underline{m}$, $\underline{n}$ inside of the symbols of differentiation; but for homogeneous solids, we treat $\underline{m}$ and $\underline{n}$ as constant. I forgot to say that δ is the cubic dilatation or the augmentation of volume per unit volume in the neighborhood of the point x, y, z, which is pretty well known, and helps us to see the relations to rigidity, and so on. If we suppose zero rigidity $P = m\delta$ is the relation between pressure and volume. In order to verify this take the second equation in P and make $n = 0$ and we obtain $P = m\delta$, the equation for the compression of a compressible fluid, in which $\underline{m}$ has become the bulk modulus.

This sort of work is called molar dynamics. It is the dynamics of continuous matter; there are no molecules, no heterogeneousnesses at all. We are preparing the way for dealing with heterogeneousnesses in the most analytical manner by supposing m and n to be functions of x, y, z, Lord Rayleigh studied the blue sky in that way, and very beautifully the treatment is quite perfect of its kind. He considers an imbeded point to represent the particle of water or dust or unknown material, whatever it is, that causes the blue sky. He supposes a sudden change of rigidity and of density in the luminiferous ether; not an absolutely sudden change, but a change not homogeneous all around, and confined to a space which is small in comparison with the wave length.

I want to take up another subject which will prepare the way to what we shall be doing afterward, which is the particular dynamical problem of the movement of a system of connected particles. I suppose most of you know the linear equations of motion of a connected system – the cycloidal motion; the equations whose integral always leads to the same formula as the cycloidal pendulum, viz: a determinant equated to zero, whose roots are essentially real.

As an example, take three weights, one of 7 pounds, another of 14 pounds, and another of 28 pounds, say. The lowest weight is hung upon the middle weight by a spiral

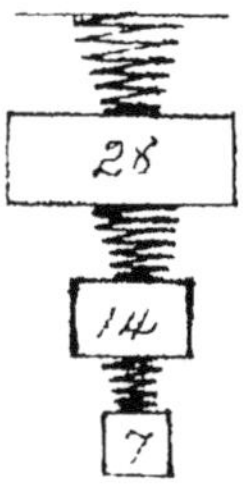

spring; the middle is hung upon the upper by a spiral spring, and the upper is attached to a fixed point by a spiral spring. It is a pretty illustration; and I find it very useful to myself. I am speaking, so to say, to Professors who sympathize with me, and might like to know an experiment which will be instructive to to their pupils.

Just apply your finger to any one of the weights – the upper weight, for example. You soon learn the periods. Move it up and down gently in the period which you find to be that of the three all moving in the same direction. You will get a very pretty oscillation, the lowest weight moving through the greatest amplitude, the second through a less, and the upper weight through the smallest. That is No 1 motion, corresponding to the greatest root of the cubic equation which expresses the solution of the mathematical problem. No. 2 motion will come after a little practice. You soon learn to give an excitation a good deal further than before, in which the lowest weight moves downward while the two upper move upwards, or the two lower move downwards while the upper moves upwards, or it might be that the middle weight does not move at all in this second mode, in which case the excitation must be by putting the finger on the upper or the lower weight. These periods depend upon the magnitude of the weights, and the strength of the springs that we use; and are soon learned in any particular set of weights and springs. It might be a good problem for junior laboratory students to find the weights and springs which will insure a case of the nodal point lying between the upper and middle weights, or at the middle weight or between the middle and lower weights. The next mode of vibration, corresponding to the smallest root of the cubic equation is one in which you always have one node between the upper and middle weights, and one node between the middle and lowest (the first and third weight vibrating in the same direction, and the middle weight in an opposite direction to the first and third)

It is assumed that there is no mass in the springs. If you want to vary your laboratory excercises, take smaller masses for the weights, and more massive springs and you pass on again to a very beautiful illustration of the velocity of sound. For that purpose a long spiral spring of steel wire 20 feet long, hung up, will answer. You can get the gravest fundamental modes without any attached weights at all. In this problem which we have been considering we have three separate weights and not a continuous spring; and we have three, and only three modes of vibration, when the springs are massless. We have an infinite number of modes when the mass of the springs is taken into account. In any convenient arrangement of heavy weights, the stiffness of the springs is so great and their masses so small that the period of vibrations of one of the springs will be very short; but take a long spring a spiral of best piano-fort steel wire, perhaps and hang it up and you will find it a nice illustration for getting the gravest fundamental modes.

I want to put down the dynamics of our problem for any number of masses. You will see at once that that is just the case that I spoke of yesterday of extending Helmholtz's singly vibrating particle connected with the luminiferous ether to a multiple vibrating heavy elastic atom imbeded in the luminiferous ether, which I think must be the true state of the case. A solid mass must act relatively to the luminiferous ether as an elastic body imbeded in it of enormous mass compared with the mas of the luminiferous ether that it displaces. In order that the vibrations of luminiferous ether may not be absolutely stopped by the mass, there must be an elastic connection. It is easier to say what must be than to say that we can understand the result. The result is almost infinitely difficult to understand in the case of ether in glass or water or carbon disulphide. but the luminiferous ether in air is very easily understood. We just think of the molecules of oxygen and nitrogen as if they were groups of jelly relative to the luminiferous ether, and you do not in the slightest degree need to take into account the motions of the particles of oxygen, nitrogen and carbon dioxide in our atmosphere relatively to the propagation of waves through the air. Think of it in this way: the period of vibration is from the 100 million millionths of a second to the 1600 million millionth of a second. Now think how far a particle of oxygen or nitrogen moves in the course of that exceedingly small time. You will find that it moves through an exceedingly small fraction of the wave length. Inasmuch as each particle moves through a very small fraction of the wave length during its period, I am fully confident that the wave motion takes place independently of the translatory motions of the particles of oxygen and nitrogen in performing their functions according to the kinetic theory of gasses. You may therefore really look upon the motion of light waves through our atmosphere as being solved by a dynamical problem such as this, applied to a case in which there is so little effective inertia, that the velocity of light is not altered, perhaps more than one-third per cent by it. More difficulties surround the subject when you come to impact on solid bodies.

In this case, let the particles of the bodies be represented by m_1, m_2 ..... m_j. I am going to suppose the several particles to be acted upon by connecting springs. I do not want to use spiral springs here. The spiral of the spring in these experiments has no effect; but I want to introduce a spiral for investigating the dynamics of the helical properties, as shown by sugar. It is usually called the rotary property, although a misnomer. The magneto-optical property which was discovered by Farraday is rotational; the property exhibited by quartz and sugar and such things has not the essential elements of rotation in it, but has the characteristic of a spiral spring in the constitution of the matter that exhibits it. We apply the word helical to the one and the word rotational to the other.

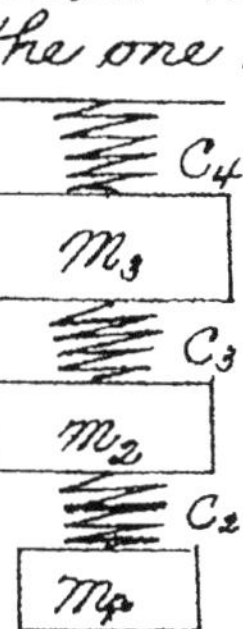

I am going to suppose one more connecting particle P, — a particle of the elastic solid, — which is moved to and fro with a given motion whose displacement downwards from a fixed point O, we shall call ξ. Let C_1 be the coefficient of elasticity of the first spring connecting the particle P with the particle m_1, C_2 the coefficient of elasticity of the next spring connecting m_1 and m_2; C_j+1, the coefficient of elasticity of the spring connecting m_j to a fixed point. We are not taking gravity into account; we have nothing to do with it. Although in the experiment it is convenient to use gravity, it would be still better if we could go to the centre of the earth and perform the experiment. The only difference would be, these springs would not be pulled out by the weights hung upon them. In all other respects the problem would be the same, and the same symbols would apply.

We are reckoning displacements downward as positive, the displacement of the particle m_i being x_i. The force acting upon m_1 in virtue of the spring connection between it and P is $C_1(\xi - x_1)$; and in virtue of the spring connection between it and m_2, is the opposing pull $-C_2(x_1 - x_2)$; so that the equation of motion of the first particle is $m_1 \frac{d^2x_1}{dt^2} = C_1(\xi - x_1) - C_2(x_1 - x_2)$. For No. 2 particle we have

$$m_2 \frac{d^2x_2}{dt^2} = C_2(x_1 - x_2) - C_3(x_2 - x_3); \text{ and so on.}$$

Now suppose P to be arbitrarily kept in some simple harmonic motion in time or period T. I might introduce a fresh set of letters and say, let $\xi = \text{Const} \times \text{Cos}\, \omega t$, ω being the angular velocity; but we take the formula $\xi = \text{Const} \times \text{Cos}\, \frac{2\pi t}{T}$. We assume that every part of the apparatus is moving with a simple harmonic motion, as will be the case if there be infinitessimal resistance and the simple harmonic motion of P is kept up long enough; so that we can write $x_1 = \text{Const}\, \text{Cos}\, \frac{2\pi t}{T}$, etc. I am going to alter the m^s so as to do away with the $4\pi^2$ which comes in from differentiation. I will let $\frac{m_1}{4\pi^2}$ denote the mass of the first particle, and $\frac{m_2}{4\pi^2}$ the mass of the second particle, etc. The result will be that the equations of motion become, $-\frac{m_1}{T^2} x_1 = C_1(\xi - x_1) - C_2(x_1 - x_2)$ etc.

Our problem is reduced now to one of algebra. There are some interesting considerations connected with the determinant which we shall obtain by elimination from these equations. To find the number of terms is easy enough; and it will lead to some remarkable expressions. But I wish particularly to treat it with a view to obtaining by very short arithmetic the result which can be obtained from the determinant in the regular way only by enormous calculation. We shall obtain an approximation, to the accuracy of which there is no limit if you push it far enough that will be exceedingly convenient in performing the calculations.

In the next lecture we shall begin with the solution of the equations that are on the board for sound. We shall then try to go on a step further with this dynamical problem.

Lecture III.

We will now go on with the problem of molar dynamics, the propagation of sound or of light, from a source. I advise you all who are engaged in teaching or in thinking of these things for yourselves, to make little models. If you want to imagine the strains that were spoken of yesterday, get such a box as this covered with white paper and mark upon it the directions of the forces S, T, U. I always take the directions of the axes in a certain order so that the direction of positive rotation shall be from y to z, from z to x, from x to y. What we call positive is the same direction as the revolution of a planet seen from the Northern hemisphere, or opposite to the motion of the hands of a watch. I have got this box for another purpose, as a mechanical model of an elastic solid with 21 independent moduluses, the possibility of which used to be disproved, and after having been proved, the result has been doubted for a long time.

Let us take our equations.

$$\rho \frac{d^2\xi}{dt^2} = \frac{dP}{dx} + \frac{dU}{dy} + \frac{dT}{dz}, \quad \text{where}$$

$$P = (m-n)\delta + 2n\frac{d\xi}{dx}, \quad U = n\left(\frac{d\xi}{dy} + \frac{d\eta}{dx}\right), \quad T = n\left(\frac{d\zeta}{dx} + \frac{d\xi}{dz}\right).$$

$$\left\{\delta = \frac{d\xi}{dx} + \frac{d\eta}{dy} + \frac{d\zeta}{dz}\right\}$$

We shall not suppose that m and n are variables, but take them constant. If we do not take them constant we shall be ready for Lord Rayleigh's paper, already referred to. I will do the work upon the board in full, as it is a case in which the expenditure of chalk saves brain; but it would be a waste to print such calculations, for the reason that a reader of mathematics should have pencil and paper beside him to work the thing out. ※ ※ ※

The result is that $\rho \frac{d^2\xi}{dt^2} = m \frac{d\delta}{dx} + n\nabla^2 \xi$ (1)

We take the symbol $\nabla^2 = \frac{d^2}{dx^2} + \frac{d^2}{dy^2} + \frac{d^2}{dz^2}$. In the case of no rigidity, or $n = 0$, the last term goes out. We shall take solutions of these equations, irrespectively of the question of whether we are going to make $n = 0$ or not, and we shall find that one standard solution for an elastic solid is independent of n and is therefore a proper solution for an elastic fluid.

I have, in this Royal Institute lecture of mine in Feb. 1883, on the Size of Atoms, inserted a note on some mathematical problems which I set when I was examiner for the Smith's Prize at Cambridge, Jan. 30, 1883. One was to show that the equations of motion of an isotropic elastic solid are what we have here obtained, and another to show that so and so was a solution. We will just take that, which is: Show that every possible solution of these three equations [(1) etc.] is included in the following:

$\xi = \frac{d\varphi}{dx} + u$, $\eta = \frac{d\varphi}{dy} + v$, $\zeta = \frac{d\varphi}{dz} + w$, where φ, u, v, w, are some functions of x, y, z, t. Of course every possible solution is included in these formulae because u, v, w, may be any functions, but the condition is added that u, v, w are such that $\frac{du}{dx} + \frac{dv}{dy} + \frac{dw}{dz} = 0$.

If we calculate the value of the cubic dilitation, we find $\delta = \nabla^2\varphi + \frac{du}{dx} + \frac{dv}{dy} + \frac{dw}{dz} = \nabla^2\varphi$.

Again, by substituting $\xi = \frac{d\varphi}{dx} + u$, in (1), we find (bearing in mind $\delta = \nabla^2\varphi$)

$$\rho\left(\frac{d^2}{dt^2}\frac{d\Phi}{dx}+\frac{d^2u}{dt^2}\right)=(m+n)\nabla^2\frac{d\Phi}{dx}+n\nabla^2u.$$

Now, we have, $\rho\frac{d^2}{dt^2}\frac{d\Phi}{dx}=(m+n)\nabla^2\frac{d\Phi}{dx}$. This is not proved as yet; the proof is reserved. Multiply this by dx, and the similar equations by dy, dz, and add. We thus get a complete differential; in other words, the relation which Φ must satisfy is $\rho\frac{d^2\Phi}{dt^2}=(m+n)\nabla^2\Phi$ [in addition to the relation $\delta=\nabla^2\Phi$, which, as will be seen in the next lecture, determines Φ as the potential corresponding to the density $-\frac{\delta}{4\pi}$]. Therefore, if Φ satisfies this, we have u, v, w, satisfying equations of the same form:

$$\rho\frac{d^2u}{dt^2}=n\nabla^2u,\quad \rho\frac{d^2v}{dt^2}=n\nabla^2v,\quad \rho\frac{d^2w}{dt^2}=n\nabla^2w.$$

By solving these four similar equations, one involving $(m+n)$, and three involving n, we can get solutions of (1), that is certain. That we get every possible solution, I shall hope to prove to-morrow. The velocity of the sound wave, or condensational wave is $\sqrt{\frac{m+n}{\rho}}$. The velocity of the wave of distortion in the elastic solid is $\sqrt{\frac{n}{\rho}}$. I shall not take this up because I am very anxious to get on with the molecular problem; but you see brought out perfectly well the two modes of waves in an isotropic homogeneous solid, the condensational wave and the distortional wave. The condensational wave follows the equations of motion of sound, which is the same as if n were null; and this gives the solution of the propagation of sound in a homogeneous medium, like air, etc. The solution is worked out ready at hand for the distortional wave because the same forms of equations give us separate components u, v, w; the same solution that gives us the velocity potential for the condensational waves, gives us the separate components of displacement for the distortional waves.

What I am going to give you to-morrow will include a solution which is alluded to by Lord Rayleigh. There is nothing new in it; Lord Rayleigh knew it perfectly well. I am going to pass over the parts of the solution which interpreted by Stokes explains that beautiful and curious experiment of Leslie's. Lord Rayleigh quotes from Stokes, ending his quotation of 8 pages with "The importance of the subject and the masterly manner in which it has been treated by Prof. Stokes will probably be thought sufficient to justify this long quotation." I would just like to read two or three things in it. Lord Rayleigh says (Theory of Sound, Vol. II, p. 207) "Prof. Stokes has applied this solution to the explanation of a remarkable experiment by Leslie, according to which it appeared that the sound of a bell vibrating in a partially exhausted receiver is diminished by the introduction of hydrogen. This paradoxical phenomenon has its origin in the augmented wave length due to the addition of hydrogen in consequence of which the bell loses its hold (so to speak) on the surrounding gas." I do not like the words "paradoxical phenomenon." "curious phenomenon", or "interesting phenomenon" would be better. There are no paradoxes in science. We may call it a dynamox, but not a paradox. Lord Rayleigh goes on to say, "The general explanation cannot be better given than in the words of Prof. Stokes: 'Suppose a person to move his hand to and fro through a small space. The motion which is occasioned in the air is almost exactly the same as it would have been if the air had been an incompressible fluid. There is a mere local reciprocating motion in which the air immediately in front is pushed forward and that immediately behind impelled after the moving body while in the anterior space generally the air recedes from the encroachment of the moving body, and in the posterior space generally flows in from all sides to supply the vacuum that tends to be created, so that in lateral directions the flow of the fluid is backwards, a portion of the excess of the fluid in front going to supply the deficiency behind.'" — It will take some careful thought to follow it. I wish I had Green here to read a sentence of his. Green says, "I have no faith in speculations of this kind unless they can be reduced to regular analysis." Stokes speculates in a way, but is not satisfied without reducing it to regular analysis. He gives here some very elaborate calculations that are also important and interesting in themselves, partly in connection with spherical harmonics, and partly from their exceeding instructiveness in respect to many problems regarding sound. Passing by all that 5 or 6 pages of mathematics — I will not tax your brains with trying to understand the dynamics of it in the course of a few minutes; I am rather calling your attention to a thing to be read than reading it — Stokes comes more particularly to Leslie's experiments. Instead of a bell vibrating, Stokes considers the vibrations of a sphere becoming alternately prolate and oblate; and he ~~considers the vibrations of a sphere,~~ shows that the principles are the same.

I have intended merely to arouse an interest in the subject. I proposed the springs as offering a solution. For any one of the springs let there be a certain change of pull C, per unit change of length. It is not the slightest matter whether a spring is long or short, only, if it is long, let it be so much the stiffer; but long or short, thick or thin, it must be massless. I mean that it shall have no inertia. I am going to put a little memorandum on the board to keep this proposed explanation by the springs in mind. I hope we will reach it today. I think it has its applications straight away to anomalous dispersion and possibly elsewhere though we are getting into the almost hopeless problem of explaining double refraction in crystals, and so on, by the wave theory of light.

To return to the consideration of these springs, we will suppose a good fixing at the top, so firm and stiff that the changing pull of the spring does not give it any sensible motion. The masses may be equal or unequal, and are connected by springs, Let us attach here a bell pull or something or other, that you can pull by, and call that

P. This, in our application to the luminiferous ether, will be the rigid shell lining between the luminiferous ether and the first moving mass.

The equation of motion for the first mass becomes, on bringing ξ to the left hand side. $-c_1\xi = \left(\frac{m_1}{q^2} - c_1 - c_2\right) x_1 + c_2 x_2$; and similarly for the second mass, I shall use i to denote any integer; I find the letter i too useful for that purpose to give it up, and when I want to write the imaginary $\sqrt{-1}$, I use $\underline{i}$. Let us call the first coefficient on the right a_1, the similar coefficient in the next equation a_2, and so on, so that $a_i = \frac{m_i}{q^2} - c_i - c_{i+1}$. The ith equation will thus be $-c_i x_{i-1} = a_i x_i + c_{i+1} x_{i+1}$. Now write down all these j equations; form the determinant by which you find all of the others in terms of ξ, and the problem is solved.

If we had a little more time I would like to determine the number of terms in this determinant. We will come back to that because it is exceedingly interesting; but I want at once to put the equations in an interesting form, borrowing a suggestion from Laplace's treatment of the celebrated Diophantine problem. What we want is really the ratios of the displacements, and we shall therefore write $\frac{c_i x_{i-1}}{-x_i} = u_i$ introducing the minus sign, so that when the displacements are alternately positive and negative the successive ratios will be all positive. We have then:

$$\frac{c_1\xi}{-x_1} = u_1 = a_1 - \frac{c_2^2}{u_2}, \quad u_2 = a_2 - \frac{c_3^2}{u_3}, \ldots \quad u_i = a_i - \frac{c_{i+1}^2}{u_{i+1}}, \ldots \quad u_j = a_j$$

$(u_{j+1} = \infty)$. We can now form a continued fraction which, for the case that we want is rapidly convergent. If this be differentiated with respect to q^2, we find a very curious law, but I am afraid we must leave it for the present. The solution is

$$u_1 = \frac{c_1\xi}{-x_1} = a_1 - \cfrac{c_2^2}{a_2 - \cfrac{c_3^2}{a_3 - \cdots \, a_{j-1} - \cfrac{c_j^2}{a_j}}}$$

Thus if we are given the spring connections and the masses, everything is known when the period is known. If you develop this, you simply form the determinant; but the fractional form has the advantage that in the case when the masses are larger and larger, and the spring connections are not larger in proportion we get an exceedingly rapid approximation to its value by taking the successive convergents. The differential coefficient of this continued fraction with respect to the period is essentially negative, and thus we are led beautifully from root to root, and see the following conditions: First suppose we move P with very great rapidity: then when the whole has come to a periodic movement, it is necessary that P and the first particle move in opposite directions. The vibrations of the first particle is hurried up when the motion of P is of a shorter period than the shortest of the possible independent motions of the system and if you want to hurry up a particle, you shove it at the end of one range and pull it at the end of the other. You meet this principle quite often; it is well known in the construction of clock escapements. To hurry up the vibratory motion we must add to the return force of particle No. 1 by the action of the spring connected to the handle P. From looking at the thing, and learning to understand it by making the experiment if you do not understand it by brains alone, you will see that everything that I am saying is obvious. It is not satisfactory to speak of these things in general terms unless we can submit them to a rigorous analysis.

That is the configuration in which the motion of P is of a shorter period than the shortest that will give us any of the critical periods. Suppose now, the motion of P to be less rapid and less rapid; a state of things will come in which, the motion of P being slower and slower, the motion of the first particle will be ~~less~~ greater and ~~less~~ greater. That is to say, if we go on diminishing and diminishing the motion we shall find for some range of motion of P, that the motion of m, and each of the other particles will be greatly ~~decreased~~ increased relatively to the motion of P. In analytical words, if we begin with a configuration of values corresponding to T very small, and then, if we increase T, making it greater and greater we shall find an infinity will appear; we shall find $\frac{x_1}{\xi}$ will become infinite. In the first place, we begin with $u_1, u_2, \ldots u_i$ all positive — T small will cause them all to be positive as you will see. Take the differential coefficient of u_i with respect to T and it will be found to be essentially negative. In other words, if we increase T, we shall diminish u_1, u_2... . In some classes of cases not necessarily in all, u_1 will first become zero; then we get the first infinity $\frac{x_1}{\xi} = \infty$. If we diminish T a little further u_2 will become zero; diminish T a little further and u_3 will become zero. We shall go into this to-morrow; but I should like to have you know beforehand what is going to come from this kind of treatment of the subject.

Lecture IV.

We found yesterday

$$\rho \frac{d^2\xi}{dt^2} = (k+\tfrac{1}{3}n)\frac{d\delta}{dx} + n\nabla^2\xi, (m=k+\tfrac{1}{3}n);$$

and we saw that we get two solutions, which when full interpreted, correspond to two different velocities of propagation, on the assumptions that were put before you as to a condensational or a distortional wave. We will approach the subject again from the beginning, and you will see at once that the sum of these solutions express every possible solution.

In one of our solutions of yesterday, we took, instead

of ξ, η, ζ, other symbols u, v, w, which satisfied the condition, $\frac{du}{dx} + \frac{dv}{dy} + \frac{dw}{dz} = 0$. In other words, the u, v, w, of yesterday express the displacements in a case in which the dilatation or condensation is zero. Now, just try for the dilatation in any case whatever, without such restriction. That we can do as follows: Differentiate (1) with respect to x, (taking account of the constancy of $\underline{m}$ and $\underline{n}$) and the corresponding equations with respect to y and z, and add. We thus find $\rho \frac{d^2\delta}{dt^2} = (m+n)\nabla^2\delta = (k + \frac{4}{3}n)\nabla^2\delta$. This equation, you will remember, is the same as we had yesterday for φ. We shall consider solutions of this equation presently; but now remark, that whatever be the displacements, we have a dilatation corresponding to some solution of this equation. When we pass on from this equation to find ξ, η, ζ, subject to other conditions, we can look upon it in this way. Suppose, for the moment $\frac{d\varphi}{dx}, \frac{d\varphi}{dy}, \frac{d\varphi}{dz}$ to be three displacements which we may compound with the actual displacements, ξ, η, ζ, if you please. I have made no supposition, as yet, as to what φ may be. I say, let these three differential coefficients denote merely three displacements at any point x, y, z. Let us now determine φ so that δ is the dilatation corresponding to them. That is to say, let us take $\nabla^2\varphi = \delta$. We know how to find φ from this equation. It is the problem of attraction, viz. $\nabla^2\varphi = -4\pi\frac{-\delta}{4\pi}$. Therefore $\frac{-\delta}{4\pi}$ will be, in the familiar case of attraction, the density of the distribution of matter of which the potential is φ; so that we shall have, $-\varphi = \iiint \frac{\delta' dx' dy' dz'}{4\pi\sqrt{[(x-x')^2+(y-y')^2+(z-z')^2]}}$; where δ' denotes the value of δ at the point x', y', z'; and we may put in the limits of integration $-\infty$ to $+\infty$. This is the familiar expression for the potential of matter of the density δ, distributed through all space. If we have other boundary conditions, we must put those in.

For any possible solutions of equations (1) etc., we have a value of δ which is a function of x, y, z; take the above volume integral corresponding to this value of δ through all points of space $x'y'z'$, and we obtain the corresponding φ function which fulfils the condition $\nabla^2\varphi = \delta$. Now, let us compound as follows the displacements $\frac{d\varphi}{dx}$, etc., with the actual displacements: $\xi - \frac{d\varphi}{dx} = u$, $\eta - \frac{d\varphi}{dy} = v$, $\zeta - \frac{d\varphi}{dz} = w$; and remarking that we have $\frac{du}{dx} + \frac{dv}{dy} + \frac{dw}{dz} = 0$, the proposition that we proposed yesterday is established.

To obtain a solution of (1) etc., we have simply to find δ from the equation $\rho\frac{d^2\delta}{dt^2} = (m+n)\nabla^2\delta$; and u, v, w from the similar equations which we found yesterday with n in the place of $(m+n)$ * subject to the conditions $\frac{du}{dx} + \frac{dv}{dy} + \frac{dw}{dz} = 0$.

We shall take our φ solution and see how we can vary that and obtain different forms of φ solutions. We can do that for the purpose of illustrating different problems in sound, and in order to familiarize you with the wave that may exist along with the wave of distortion in any true elastic solid which is incompressible. We ignore this condensational wave in the theory of light. We are sure that its energy, at all events if it is not null, is very small in comparison with the luminiferous vibrations we are dealing with. But to say that it is absolutely null would be an assumption that we have no right to make. When we look through the little Universe that we know, and think of the transmission of electrical force and of the transmission of magnetic force and of the transmission of light, we have no right to assume that there may not be something like that our philosophy does not dream of. We have no right to assume that there may not be condensational vibration in the luminiferous ether. We only do know that any vibrations of this kind which are excited by the reflection and refraction of light are certainly of very small

---

* [This requires that φ should satisfy the same equation as δ, which is the proposition (left undemonstrated in the last lecture) by which the equations for u, v, w were obtained. Later on in response to a question raised by Dr. Franklin an indirect proof of the proposition $\rho\frac{d^2\varphi}{dt^2} = (m+n)\nabla^2\varphi$ is given. A direct demonstration may be obtained from the value of $\varphi = \iiint \delta' \frac{dx'dy'dz'}{4\pi r}$, remembering that $\rho\frac{d^2\delta}{dt^2} = (m+n)\nabla^2\delta$, and that $-\iiint \frac{\nabla'^2\delta'}{4\pi r}dx'dy'dz' = \delta = \nabla^2\varphi$. H.]

energy compared with the energy of the light from which they proceed. The fact of the case as regards reflection and refraction is this, that unless the luminiferous ether is absolutely incompressible, the reflection and refraction of light must generally give rise to waves of condensation. Waves of distortion may exist without waves of condensation, but waves of distortion cannot be reflected at the bounding surface between two mediums without exciting in each medium a wave of condensation. When we come to the subject of reflection and refraction, we shall see how to deal with these condensational waves and find how easy it is to get quit of them by supposing the medium to be incompressible. But it is always to be kept in mind to be examined into, are there or are there not very small amounts of condensational waves generated in reflection and refraction, and may after all, the law of electric force not depend on the waves of condensation.

Suppose that we have at any place in air, or in luminiferous ether (I cannot distinguish now between the two ideas) a body that through some action we need not describe, but which is conceivable, is alternately positively and negatively electrified; may it not be that this will be the cause of condensational waves? Suppose this, that we have two spherical conductors united by a fine wire, and that an alternating electromotive force is produced in that fine wire, for instance with an alternating dynamo-electric machine; and suppose that sort of thing goes on away from disturbance — at a great distance up in the air, for example. The result of the work of that dynamo-electric machine will be that one conductor will be alternately positively and negatively electrified and the other conductor negatively and positively electrified. It is perfectly certain that if we turn the machine slowly that in the neighborhood of the conductors we will have alternately positively and negatively electrified elements with reversals, perhaps two or three hundred per second of time without a gradual transition from negative through to zero, to positive, and so on; and the same thing all through space; and we can tell exactly what the potential is at each point. Now, does any one believe that if that revolution was made fast enough that the electro-static law would follow? Every one believes that if that process be conducted fast enough, several million times, or millions of million times per second we should be far from fulfilling the electrostatic law in the electrification of the air in the neighborhood. It is absolutely certain that such an action as that going on would give rise to electrical waves. Now it does seem to me probable that those electrical waves are condensational waves in luminiferous ether; and probably it would be that the propagation of these waves would be enormously faster than the propagation of ordinary light waves.

I am quite conscious, when speaking of this of what has been done in the so-called Electro-Magnetic theory of light. I know the propagation of electric impulse along an insulated wire surrounded by gutta percha, which I worked out myself, about the year 1854 and in which I found a velocity comparable with the velocity of light. We then did not know the relation between electro-static and electro-magnetic units. If we had, that might have been obtained in the way that Maxwell has brought out so beautifully from the proper coefficients of capacity for the gutta percha. If we work that out for the case of air instead of gutta percha, we get practically the same v, I think, for the velocity of propagation of the impulse. That is a very different case from this and I have waited in vain to see how we can get any justification of the way of putting it in the so-called electro-magnetic theory of light. Simplify it down to the uttermost and take that case; there is a case of excitation of a kind that we know; we know the a, b, c of it, and the laws of it, and feel certain that if this operation be performed, but fast enough there will be waves. It seems to me that there are exceedingly strong probabilities that these will be waves of condensation and rarefaction of the luminiferous ether. I may refer to a little article of mine in which I gave a sort of mechanical representation of electric, magnetic, and galvanic forces — galvanic force I called it then, a very badly chosen name. It is published in the first volume of the reprint of my papers. It is shown in that paper that the static displacement of an elastic solid follows exactly the laws of the electro-static force, and that rotary displacement of the medium follows exactly the laws of magnetic force.

It seems to me that an incorporation of the theory of the propagation of electric and magnetic disturbances with the wave theory of light is most probably to be arrived at by this view that I am now indicating. In the wave theory of light, however, we shall simply suppose the resistance to compres[illegible] of the luminiferous ether and the velocity of propagation of the condensational wave in it to be infinite. We shall sometimes use the words "practically infinite" to guard against supposing these quantities to be absolutely infinite.

I will now take two or three illustrations of this solution for condensational waves. Part of the problem that I referred to yesterday says, prove that the following is a solution of these equations $\varphi = \frac{1}{r} \sin \frac{2\pi}{\lambda}\left(r - t\sqrt{\frac{m+n}{\rho}}\right)$, $\left[=\frac{1}{r}\sin q\right]$

We might put this in a more analytical form, but the analysis consists in the verification of the thing. For that purpose, let us take the Laplacian of φ. We use this theorem, $\frac{d^2}{dx^2}(uv) = v\frac{d^2u}{dx^2} + 2\frac{du}{dx}\frac{dv}{dx} + u\frac{d^2v}{dx^2}$, and find

$$\nabla^2\varphi = \frac{1}{r}\left\{\frac{2\pi}{\lambda}\cdot\frac{2}{r}\cos q - \frac{4\pi^2}{\lambda^2}\sin q\right\} - 2\frac{2\pi}{\lambda}\frac{1}{r^2}\cos q + 0 = -\frac{4\pi^2}{\lambda^2}\varphi$$

Our equation for φ is therefore

$$\rho\frac{d^2\varphi}{dt^2} = (m+n)\nabla^2\varphi = -(m+n)\frac{4\pi^2}{\lambda^2}\varphi$$

We will now make it a little more analytical and say the thing to be proved is that which is written down letting the assumption be: $\varphi = \frac{1}{r} \sin 2\pi \left(\frac{r}{\lambda} - \frac{t}{T} \right)$, where T is the period of vibration and λ the wave length. Substitute and the equation becomes $\rho \frac{4\pi^2}{T^2} \varphi = - \frac{4\pi^2}{\lambda^2} (m+n) \varphi$; or the velocity of propagation $\frac{\lambda}{T}, = \sqrt{\frac{m+n}{\rho}} = \sqrt{\left(\frac{k+\frac{4}{3}n}{\rho}\right)}$.

There then is the determination of a form of motion which is possible for an elastic solid. We shall consider the nature of this motion presently. The presence of $\frac{1}{r}$ prevents it from being a pure wave motion. Passing over that consideration for the present, we note that it is less and less effective, relatively to the motion considered the further we go from the center.

In the meantime, we remark that the velocity of propagation in an elastic solid is little greater than in a fluid with the same resistance to compression. k is the bulk modulus and measures resistance to compression; $\underline{n}$ is the rigidity modulus. I may hereafter consider relations between $\underline{k}$ and $\underline{n}$ for real solids. k is generally several times $\underline{n}$, so that $\frac{1}{3}n$ is very small in comparison with k, and therefore in ordinary solids the velocity of propagation of the condensational wave is exceedingly little greater than if the solid were deprived of rigidity and we had an elastic fluid of the same bulk modulus.

I shall want to look at this motion in the neighborhood of the source. That beautiful investigation of Stokes quoted by Lord Rayleigh has to do entirely with the region in which the change of value of this coefficient $\left(\frac{1}{r}\right)$ from point to point is considerable. Without looking at that now, let us find the displacement and see what it will be. $\frac{d\varphi}{dx}, \frac{d\varphi}{dy}, \frac{d\varphi}{dz}$, are the three components of the

---

*[The lecturer used throughout this investigation $m = k + \frac{1}{3}n$, instead of $m+n = k + \frac{4}{3}n$. The fact that this should be $\frac{4}{3}n$ is the occasion for a further consideration of the subject in a subsequent Lecture. H.]

displacements clearly, the displacement will be in the direction of the radius because everything is symmetrical; and its magnitude will be $\frac{d\varphi}{dr}$.

Dr. Franklin:– The equation at the top of the board puzzles me. It is the same equation we have had before for δ, with φ written in the place of δ. I do not understand how the φ got there.

Sir Wm. Thomson:– Let us see how this is. It is quite correct, but we will just look at that equation. Our equation was $\rho\frac{d^2\delta}{dt^2}=(m+n)\nabla^2\delta$; and then we had $\nabla^2\varphi=\delta$. δ must fulfil the first condition, and if φ fulfils that condition δ does certainly fulfil it in virtue of the second condition. That ought to prove that when δ fulfils the first condition φ also fulfils it. That is not quite rigorous perhaps, but I think it is obvious from the finding of φ from δ. If we take the Laplacean of $\rho\frac{d^2\varphi}{dt^2}=(m+n)\nabla^2\varphi$ we find (since $\nabla^2\varphi=\delta$) $\rho\frac{d^2\delta}{dt^2}=(m+n)\nabla^2\delta$. All we have to do is to find δ to fulfil this condition, and having found it, we will find φ to fulfil it.

Having obtained a solution of our equations, let us see what we can make in interpreting it. The component of the displacement in the direction of x is $\frac{d\varphi}{dx}=\frac{2\pi x}{\lambda r^2}\times\left(\cos q-\frac{\lambda}{2\pi r}\sin q\right)$. When r is great in comparison with $\frac{\lambda}{2\pi}$, the second term becomes very small in comparison with the first and we have $\frac{d\varphi}{dx}=\frac{2\pi}{\lambda}\frac{x}{r^2}\cos q$. Also $\frac{d\varphi}{dr}=-\frac{1}{r^2}\sin q+\frac{2\pi}{\lambda}\frac{1}{r}\cos q$. Therefore, when the distance from the origin is large in comparison with $\frac{\lambda}{2\pi}$, the displacement is sensibly equal to $\frac{2\pi}{\lambda}\frac{1}{r}\cos q$, and is therefore approximately in the inverse proportion to the distance; and the intensity of the sound if it were to be applied to sound, would be inversely as the square of the distance. At a considerable distance from the place in which there is circulation around the source, that is the permanent term which I have written down.

I want to get a second and a third solution. Take $\psi = \frac{\lambda}{2\pi}^{*} \frac{d\phi}{dx} = \frac{x}{r^2}\left(\cos q - \frac{\lambda}{2\pi r}\sin q\right)$ as the velocity potential for a fresh solution. I take it that you all know that if we have one solution ϕ, for the velocity potential, we can get any other solution by any linear function of $\frac{d\phi}{dx}, \frac{d\phi}{dy}, \frac{d\phi}{dz}$. Now let us find the displacements $\frac{d\psi}{dx}, \frac{d\psi}{dy}, \frac{d\psi}{dz}$. Here I want to prove that though this solution is no longer symmetrical with respect to r, so that there will be motions other than radial in the neighborhood of the source, yet that the motion is approximately radial at a distance from the source. Work it out, and you will find that

$$\frac{d\psi}{dx} = \frac{2\pi}{\lambda}\frac{x^2}{r^3}\left[-\sin q + \frac{\lambda}{2\pi r}\cdot\frac{(r^2 - 3x^2)}{x^2}\cos q\right]$$

The principal term is then $-\frac{2\pi}{\lambda}\frac{x^2}{r^3}\sin q$. We might go on to the third and fourth terms, increasing the multiplicity. That splendid work of Stokes in which this multiplicity is dealt with to show the effect of hydrogen in killing sound is one of the finest things written in physical mathematics. But we will drop those terms and think only of the principal terms.

The principal term in the expression for the displacement is $\xi \doteqdot -\frac{2\pi}{\lambda}\frac{x^2}{r^3}\sin 2\pi\left(\frac{r}{\lambda} - \frac{t}{T}\right)$ I use $\doteqdot$ as the sign for approximate equality. This approximate equality is true for assigned distances from the centre great in comparison with the wave length. Let me remark, it is the differentiation of $\sin q$ that gives the effective terms of the displacement; and in differentiating ψ with respect to y, you have simply to take the differential coefficient of r in $\sin q$ with respect to y, instead of x. So that we may just write down the principle terms

---

* $\lambda/2\pi$ is introduced merely for convenience. The solution is not essentially different from $\psi = d\phi/dx$.

of the y and z displacements $\eta = -\frac{2\pi}{\lambda}\,\frac{xy}{r^3}\sin q$, $\zeta = -\frac{2\pi}{\lambda}\,\frac{xz}{r^3}\sin q$. These component displacements, being proportional to x, y, z, show that the resultant displacement is in the direction of the radius, and that its magnitude is $-\frac{2\pi}{\lambda}\,\frac{x}{r^2}\sin q$. If we write $x = r\cos i$, this becomes $-\frac{2\pi}{\lambda}\,\frac{\cos i}{r}\sin q$; or the displacement is inversely proportional to the distance. If $i = 0$ we have a maximum; if $i = \frac{\pi}{2}$ we have zero. The upshot of it is that the displacement is a maximum in the axis OX, zero in the axes OY, OZ, and symmetrical with respect to the axes.

The third solution is to take $\frac{d^2\phi}{dx^2}$ as our velocity potential. At a distance from the origin, great in comparison with the wave length, the displacement is in the direction of the radius, and its magnitude is $\frac{d}{dr}\,\frac{d^2\phi}{dx^2}$.

Now the interpretation of these cases is as follows:– The first solution, a globe alternately becoming larger and smaller; the second solution, a globe vibrating to and fro in a straight line; the third solution, two globes vibrating to and fro meeting one another, or the disturbance in the neighborhood of the prongs of a tuning fork.

That, however, requires a little nice consideration, and we shall take it up in a subsequent lecture. The third mode does not quite represent the motion in the neighborhood of the prongs of a tuning fork; there must be an unknown amount of the first mode compounded with the third mode for this purpose. The expression for the vibration in the neighborhood of a tuning fork, going so far from the ends of it that we will be undisturbed by the shape of the thing, will be given by the velocity potential $A\phi + \frac{d^2\phi}{dx^2}$. That will be the velocity potential for the chief terms, the terms which alone have an effect at a distance. The differentiation will be performed simply with reference to the r in the term sin q or cos q; and will be the same as if the coefficient of sin q or cos q were constant. A differentiation of this velocity potential will show that the displacement is in the direction of the radius from the centre of the system, and the magnitude of the displacement will be $\frac{d}{dr}\left(A\varphi + \frac{d^2\varphi}{dx^2}\right)$

A is an unknown quantity depending upon the tuning fork. I want to suggest this as a junior exercise, to try tuning forks with different breadths of prongs. When you take tuning forks with prongs a considerable distance assunder you have much less of the φ to take. Try a tuning fork with flat prongs, pretty close together, and you will have much more of the φ to take. The φ part of the velocity potential corresponds to the swelling of the tuning fork, the becoming larger and smaller. The larger and flatter the prongs are the greater is the proportion of the φ solution, and the larger the value of A in that formula.

The experiment that I suggest is this: That you take tuning forks and turn them around until you find the cone of silence, or find the angle between the line joining the prongs and the line going to the place where you hear no sound. The suddenness of transition from sound to no sound is startling. Having the tuning fork in the hand, turn it slowly around near one ear until you find the place of silence; a very small angle of turning around the verticle axis from that place gives you a loud sound. I think it is very likely that the place of no sound will depend on the angle of vibration. If you excite it very powerfully, you will find greater inclination; less powerfully, less inclination. It will certainly vary with the tuning fork.

Lecture V.

I stated in the last lecture that the second solution corresponding to the velocity potential $\frac{d\phi}{dx}$, would represent the effect, at a great distance from the mean position, of a body vibrating to and fro in a straight line. I said a sphere, but we may take a body of any shape vibrating to and fro in a straight line, and at a very great distance from the mean position, the motion produced will be represented by the velocity potential $\frac{d\phi}{dx}$. Then the velocity potential $\frac{d^2\phi}{dx^2}$, or the third solution, would, I believe, represent (without an additional term $A\phi$) the motion at a distance, when the origin of the sound consists in two globes, let us say, for fixing the ideas placed at a distance from one another very great in comparison with their diameters and set to vibrating to and fro. Suppose this is a globe in one hand, and this is one in the other. I now move my hands towards and from each other – that sort of motion produced by the exciting bodies would, at a very great distance be expressed exactly by the velocity potential $\frac{d^2\phi}{dx^2}$.

But when you have two globes, or two flat bodies, near one another, you need an unknown amount of the ϕ vibration to represent the actual state of the case. That unknown amount might be determined theoretically for the case of two spheres. The problem is analagous to Poisson's problem of the distribution of electricity upon two spheres, and it has been solved by Stokes for the case of fluid motion. [See Mem. de l'Inst, Paris. 1811 pp. 1, 163 & Stokes Papers, Vol. I, p. 230– "On the resistance of a fluid to two oscillating spheres"]. You can thus tell the motion exactly in the neighborhood of two spheres vibrating to and fro provided the amplitudes of their vibrations are small in comparison with the distance between them; and you can find the value of $\underline{A}$ for two spheres of any given radii and any given distance between them. For such a thing as a tuning fork, you could not, of course, work it out theoretically; but I think it would be an interesting experiment for junior laboratory work.

I suppose you are all familiar with the zero of sound in a tuning fork; but I have never seen it described correctly anywhere. I shall take that up on Monday. We shall see that we have no theoretical means of determining the inclination of the line going to the mean position of the area for silence to the line joining the prongs; but that this is dependent upon the proportions of the body. On turning the tuning fork around, you can get with great nicety the position for silence; and a surprisingly small turning of the tuning fork from the position of silence causes the motion to be heard. It would be very curious to find whether the position of zero sound varies relatively to the fork as the amplitude of the vibrations increases. I doubt whether any perceptible difference will be found in any ordinary case however we vary the amplitude of the vibrations. But I am quite sure you will find considerable difference, according as you take tuning forks of cylindrical proportions or tuning forks like the more modern ones that Koenig makes, with very broad flat ends.

Now for our molecular problem.

I want to see how the quantities vary, when we vary the period. Remember that $a_i = \frac{m_i}{T^2} - c_i - c_{i+1}$, so that $\frac{d\,a_i}{dT^{-2}} = m_i$. Write for the moment δ for $\frac{d}{dT^{-2}}$, and differentiate the equation for u_i; we have $\delta u_i = m_i + \left(\frac{c_{i+1}}{u_{i+1}}\right)^2 \delta u_{i+1}$, $\delta u_{i+1} = m_{i+1} + \left(\frac{c_{i+2}}{u_{i+2}}\right)^2 \delta u_{i+2}, \ldots \delta u_j = m_j$.

Substitute successively, and we find,

$$\delta u_i = m_i + \left(\frac{c_{i+1}}{c_{i+1}}\right)^2 m_{i+1} + \left(\frac{c_{i+1}\,c_{i+2}}{u_{i+1}\,u_{i+2}}\right)^2 m_{i+2} + \cdots \left(\frac{c_{i+1}\cdots c_j}{u_{i+1}\cdots u_j}\right)^2 m_j$$

This is our expression, and remark the exceedingly important property of it that it is essentially positive, i.e., the variation of u_i with respect to T^{-2} is essentially positive. Also $\frac{du_i}{dT} = 2T^{-3}\,\delta u_i$. Now, $u_i = \frac{c_i\,x_{i-1}}{-x_i}$, or $\frac{c_i}{u_i} = -\frac{x_i}{x_{i-1}}$, $\frac{c_i\,c_{i+1}}{u_i\,u_{i+1}} = \frac{x_i\,x_{i+1}}{x_{i-1}\,x_i}$, etc. The result therefore is this remarkable expression for the differential coefficient of u_i with respect to the period.

$$\frac{du_i}{dT} = -\frac{2}{T^3}\cdot\frac{1}{x_i^2}\left(m_i x_i^2 + m_{i+1} x_{i+1}^2 + \cdots m_j x_j^2\right) \ldots \quad (2)$$

This is certainly a very remarkable theorem, and one of great importance with reference to the interpretation of the solution of our problem. Remember that x_i is the displacement of m_i at any part of the motion. You may habitually think of the maximum values of the displacements but it is not necessary to confine yourselves to the maximum values. Instead of $x_1, x_2, \ldots x_j$ we may take constants equal to the maximum values of the x's, multiplied into $\sin\frac{2\pi t}{T}$ — remembering that each of them varies with a simple harmonic motion. The masses are positive, and we have squares of the displacements, so that the second member of (2) is essentially negative. Hence, as we augment the period, the functions u_i, etc., each one decreases and as we decrease the period, each one increases.

Let us now consider this spring arrangement. I am going to suppose, in the first place that the period of vibration is very small, and is then gradually increased. As you increase the period, the values of each one of the quantities $u_1, u_2, \ldots$ decreases. It is interesting to remark that since $\frac{du_i}{dT}$ is always negative, every one of the u's decreases throughout every variety of configuration, as T increases. In the first place, T may be taken so small that the u's are all very large positive quantities; for $u_i = \frac{m_i}{T^2} - c_i - c_{i+1} - \frac{c_{i+1}^2}{u_{i+1}}$ may be certainly made very large positive by taking T small enough if, at the same time the succeeding quantity, u_{i+1}, is large (a condition which is fulfilled since we always have $u_{j+1} = \infty$.)

Observe that the u's all positive implies that $\xi, x_1, \ldots x_j$ are alternately positive and negative. In other words the handle P and the several particles $m_1, m_2, \ldots m_j$, are each moving in a direction opposite to its neighbor. Since the magnitudes of the ratios $u_1, u_2, \ldots u_j$ of the several amplitudes, decrease with the increase of the period, the amplitude of particle m_i is becoming smaller in proportion to the amplitude of the succeeding particle m_{i+1} that is to say, the handle P is hurrying up the system. I am going to show you that as every one of these quantities u_i decreases, the first that passes through zero is u_1 — corresponding to infinite motion of the particles of the system, in comparison with the motion of the handle P. This is the first critical case; after that u_1 becomes negative, and the motion of P is in the direction of the motion of the first particle. Also, if we have farther negative values, the order of procedure always is that the negative value passes along the line from particle m_1 towards the fixed end. In other words, as we go on increasing the period we shall find that the next critical case that takes place is that particle m_1 has zero motion, or $u_1 = \frac{c_1 \xi}{-x} = -\infty$.

Let us look at the state of things when u_i has approached very near to zero. We shall have $u_{i-1} = a_{i-1} - \frac{c_i^2}{u_i}$ a

very large negative quantity. This fact alone shows that u_{j-1} must have preceded u_j in becoming zero, since it must have passed through zero before becoming large negative.*
Therefore, as we augment T, the first of the u's to become zero is $u_1 = \frac{c_1 \xi}{-x_1}$; or the motion of particle M_1 and also of each of the other particles is infinite in comparison with the motion of P. Just before this state of things all the particles $P, M_1, \ldots M_j$ are moving each opposite to its neighbor; just after it, P has reversed its motion with reference to the first particle, and is moving in the same direction with it.

That is also the configuration, just before the second critical case, in which we have u_1 large negative, u_2 small positive $u_3, \ldots u_j$, all positive. At this critical case, we have $u_1 = \frac{c_1 \xi}{x_1} = -\infty$ or $x_1 = 0$. The period of motion of P that will produce this state of things is equal to the period of the free vibration of the system of particles, with mass M_1 held at rest and each of the other masses moving in an opposite direction to its neighbor. When the period of P is equal to the period of motion of the system with the first particle held at rest, then the only motion of the system that fulfils the condition of being a simple harmonic motion is that in which the amplitude of vibration of the second particle in one direction is such as to produce a pull in that direction, equal to the pull exercised on the spring by P in the opposite directions; which keeps the first particle at rest. Immediately after this critical case, u_1 has changed from large negative to large positive and u_2 from small positive to small negative; or the first particle has reversed the direction of its motion with respect to P and the second particle.

The third critical case is that of the second particle coming to rest, and reversing its motion; but I shall not go further with these critical cases. ¶

*[This does not show, however, that u_{j-1} may not have passed through zero more than once before u_j. H.].

am only giving you an indication of how to perceive the thing. There is a great deal more to think of, as to the a's becoming negative, etc., My object was simply to indicate the state of things merely, and I will just jump over the remaining critical cases, and take up T very great.

It would be curious to find the solution when the period is infinitely great out of these equations. When T is infinite, $\frac{m_i}{T^2}$ vanishes, and $a_i = -c_i - c_{i+1}$. That applied to the equations for the u's ought to find the solutions quite readily. The solutions which you find are very curious, but it is like the case of so many problems which all the great mathematicians used to be fond of proposing and if putting their heads together to solve. If you were successfull in finding out the right way of doing them the solutions were easy, otherwise they were hard.

You know, when you think of this case, that when T is infinitely great, P is moving infinitely slowly, so that the inertia of each particle has no sensible effect and all the particles are in equilibrium. Let F be the force, then, on the spring, that is to say, pull P down with a force F and hold it at rest. What will be the displacements of the different particles? Answer, $x_j = \frac{F}{c_{j+1}}$, $x_{j-1} = \frac{F}{c_{j+1}} + \frac{F}{c_j}$ and so on. The number jth particle is displaced to a distance equal to the force divided by the coefficient of elongation of the spring. To obtain the displacement of particle $j-1$, we have to add the displacement resulting from the elongation of the next spring c_j, and so on. The general equation then is $x_i = \left(\frac{1}{c_{j+1}} + \frac{1}{c_j} + \cdots \frac{1}{c_{i+1}}\right) F$.

$$u_i = -c_i \left(\frac{1}{c_{j+1}} + \cdots \frac{1}{c_i}\right) \Big/ \left(\frac{1}{c_{j+1}} + \cdots \frac{1}{c_{i+1}}\right).$$

It is a curious problem to substitute the value of $a_i = -c_i - c_{i+1}$ in the continued fraction which gives u_i, and verify this solution.

I just want to call your attention a little bit to magnitudes; for the problem we really care for is not this. It is like fiddling while Rome is burning to be explaining the fluorescence when the explanation of the refraction of light in crystals is waiting. The difficulty is not to explain phosphorescence and fluorescence but to explain why there is so little sensible fluorescence and phosphorescence. This thing brings everything to fluorescence and phosphorescence. The state of things as regards our system would be this: Suppose we have this handle P moved backwards and forwards until everything is in a perfectly periodic state. Then suddenly stop moving P. The system will continue vibrating for a definite time with a complex vibration which will really embody something of all the modes. That I believe is fluorescence.

But now comes Mr. Michelson's question, and Mr. Newcomb's question, and Lord Rayleigh's question, as to velocity of groups. There again we are all afloat with vibrations of this kind. Suppose a succession of luminiferous vibrations commences. In the commencement of the luminiferous vibrations the attached molecules imbedded in the luminiferous ether, do not immediately get into the state of a simple harmonic vibration which will create a regular light. It seems quite certain that there must be an initial fluorescence. Let light begin shining on uranium glass; for the thousandth of a second, perhaps, after the light has begun shining on it, you should find an initial state of things, which differs from the permanent state of things exactly the same as fluorescence differs from no light at all.

There is still another question, which is of profound interest, and seems to present many difficulties, and that is, the actual condition of the light which is a succession of groups. Lord Rayleigh has told us in his printed paper in respect to the agitated question of the velocity of light, and then again at the meeting of the British Association at Montreal he repeated very peremptorily and clearly the fact that the velocity of a group of waves must not be confounded with the wave velocity of an infinite succession of waves. He seems to be quite certain that what he said is true. But here is a difficulty which has only occurred to me since I began speaking to you on the subject; and I hope, before we separate, we shall see our way through it. All light consists in a succession of groups. Why is light not polarized? We are going to work our way slowly on until we get expressions for sequences of vibrations of existing light. Take any conceivable supposition as to the origin of light, in a flame, or a wire made incandescent by an electric current or any other source of light; we shall work our way up from these sound equations to the kinds of expression that light must have from any conceivable source. Now, in a source consisting of a motion that kept going on in exactly the same way, the light from that source would be plane polarized, or circularly polarized, or elliptically polarized and would be absolutely constant. In reality, there is a multiplicity of succession of groups of waves. One molecule, of enormous mass in comparison with the luminiferous ether that it displaces gets a shock and it performs a set of vibrations until it comes to rest or gets a shock in some other direction; and it is sending forth vibrations with the same want of regularity that is exhibited in a group of sounding bodies consisting of bells, tuning forks, organs, etc., every one of which is sending forth its strain and each of which is propagated, some distance away from the source, as if there were no others. We thus see that light is entirely compounded of groups of waves, and if the velocity of a group of waves, or even the center of gravity of a group, differs from the velocity of absolutely continuous sequences of waves, we have all ground cut from under us in respect to the velocity of waves of light.

I mean to say, that all light consists of groups following one another in that way, and that there is a difficulty to see what to make of the beginning and end of the vibrations of a group. And that then there is the question which was talked over a little in Section A at Montreal, will the mean effect of the group be the same as that of an infinite sequence of uniform waves, and will the deviation from regular periodicity at the beginning and end of the group have but a small influence in comparison with the whole. It seems almost certain that it must have but a small influence from the known facts regarding the velocity of light and the approximate regularity that we have — but I am leading you into a muddle because I am far into the difficulty and have not understood it. Still you can all think a good deal with me about the connections of this subject.

Lecture VI.

I want to ask you to note that when I spoke of $k + \frac{4}{3}n$ not differing scarcely from k for most solids, I was rather under the impression for the moment that the ratio of n to k was smaller than it is; and also you will remember that we had $k + \frac{1}{3}n$ on the board. The square of the velocity of a condensational wave in an elastic solid is $k + \frac{4}{3}n$. For solids fulfilling the supposed relation of Navier and Poisson between compressibility and rigidity we have $n = \frac{3}{5}k$; and for such cases the numerator becomes $\frac{9}{5}k$. It would be k if there were no rigidity; it is $\frac{9}{5}k$ if the rigidity is that of a solid for which Poisson's ratio has its supposed value.

Metals are not enormously far from fulfilling this condition but it seems that for elastic solids generally, n bears a less proportion to k than this. It is by means certain that it fulfils it even approximately for metals; and for india rubber on the other hand, and for jellies n is an exceedingly small fraction of k, so that in these cases the velocity of the condensational wave is $\sqrt{\frac{k}{\rho}}$. The velocity of propagation of a distortional wave is $\sqrt{\frac{n}{\rho}}$; so that for jellies, the velocity of propagation of condensational wave is enormously greater than that of distortional waves.

I am asked to define velocity potential. Those who have read German writers on Hydrodynamics already know the meaning of it perfectly well. It is purely a technical expression which has nothing to do with potential or force. "Velocity potential is a function of the coördinates such that its rate of variation per unit distance in any direction is equal to the component of velocity in that direction. A velocity potential exists when the distribution of velocity are expressible in this way, in other words when the motion is an irrotational one. The most convenient definition of irrotational motion is, the motion such that the velocity components are expressed by the differential coefficients of a function. That function is the velocity potential. When the motion is rotational there is no velocity potential.

This is the strict application of the words "velocity potential" which I have used. A corresponding language may be used for displacement potential. It is not good language, but it is convenient, it is rough and ready. So that when we are speaking of component displacements in any case, whether of static displacement in an elastic solid or of vibrations, in which the components of displacement are expressible as the differential coefficients of a function, we may say that it is an irrotational displacement. If from the differentiation of a function we obtain components of velocity, we have velocity potential; whereas, if we get components of displacement we have displacement potential. The functions φ, that we used are not then, strictly speaking, velocity potentials but displacement potentials.

I want you in the first place to remark what is perfectly well known to all who are familiar with Differential equations, that taking the solution $\varphi = \frac{1}{r} \cdot \sin q$ as a primary (where $q = \frac{2\pi}{\lambda}\left(r - t\sqrt{\frac{n}{\rho}}\right)$ if we are considering the distortional wave) we may derive other solutions by differentiaticitions with respect to the rectangular coördinates. The first thing I am going to call attention to is that at a distance from the origin, whatever be the solution derived from this primary by differentiation, the corresponding displacement is nearly in the direction through the origin of coordinates.

Take any differential coefficient whatever, $\frac{d^{i+j+k}\varphi}{dx^i\,dy^j\,dz^k}$; the term of this which alone is sensible at an infinitely great distance is that which is obtained by successive differentiation of $\sin q$. That distance term in every case is as follows: $\left(\frac{2\pi}{\lambda}\right)^{i+j+k}\left(\frac{dr}{dx}\right)^i\left(\frac{dr}{dy}\right)^j\left(\frac{dr}{dz}\right)^k \frac{1}{r}\,{}^{\sin}_{\cos}\,q$. It will be $\sin q$ or $\cos q$, according as $i+j+k$ is even or odd. We do not need to trouble ourselves about the algebraic sign, because we shall make it positive, whether the differential coefficient is positive or negative. Now $\frac{dr}{dx} = \frac{x}{r}$, $\frac{dr}{dy} = \frac{y}{r}$, $\frac{dr}{dz} = \frac{z}{r}$. Thus our type solution becomes, omitting the constant factor, $\frac{x^i y^j z^k}{r^{i+j+k+1}}\,{}^{\sin}_{\cos}\,q$. This expresses the most general type of displacement potential for a condensational wave proceeding from a center. I have not formally proved that this is the most general type, but is very easy to do so. I am rather going into the thing synthetically. It is so thoroughly treated analytically by many writers that it would be a waste of your time to go into anything more, at present, than a sketch of the manner of treatment, and to give some illustrations.

But now to prove that the displacement at a distance from the origin of the disturbance is always in the direction of the radius vector. Once more, the differential coefficient of this displacement potential, which has several terms depending upon the differentiation of the r's x's, etc. has one term of paramount importance, and that is the one in which you get $\frac{2\pi}{\lambda}$ as a factor. The smallness of λ in proportion to the other quantities makes the factor $\frac{2\pi}{\lambda}$ give importance to the term in which it is found. The distance terms then for the components of the displacement are $\xi = \frac{x^i y^j z^k}{r^{i+j+k+1}} \cdot \frac{2\pi}{\lambda} \cdot \frac{x}{r} \frac{\cos}{\sin} q = R\frac{x}{r}$; $\eta = R\frac{y}{r}$, $\zeta = R\frac{z}{r}$. These are then the components of a displacement which is radial; and the expression for the radial displacement is $R = \frac{2\pi}{\lambda} \cdot \frac{x^i y^j z^k}{r^{i+j+k+1}} \frac{\cos}{\sin} q$.

The sum of any number of such expressions will express the distance effect of sound proceeding from a source. It is interesting to see how, simply by making up an algebraic function in the numerator out of the x's, y's and z's, we can get a formula that will express any amount of nodal subdivision where silence is felt. The most general result for the radial displacement is $R = \Sigma \frac{c x^i y^j z^k}{r^{i+j+k+1}} \frac{\cos}{\sin} q$. Remark that $\frac{x}{r}, \frac{y}{r}, \frac{z}{r}$ are merely angular functions and may be expressed at once as $\sin\theta \times \cos\psi$, $\sin\theta \sin\psi$, $\cos\theta$; and that therefore R is an integral algebraic function of $\sin\theta\cos\psi$, $\sin\theta\sin\psi$, $\cos\theta$.* It

---

* [The Lecturer had not the factor $\frac{\cos}{\sin} q$ upon the board in his expression for R, and so overlooked that the factor $\frac{1}{r}\cos q$ must enter into terms of even order, and $\frac{1}{r}\sin q$, into terms of odd order. Thus the most general function gives $R = \frac{1}{r}(R_0 \cos q + R_1 \sin q)$; and we have merely lines of silence radiating from the source viz., the intersections of the cones $R_0 = 0$, $R_1 = 0$. For cones of silence, either R_0, R_1 may have a common factor or one of these angular functions may be wanting in the expression for R. H.]

is thus easy to see that you can vary indefinitely the expressions for sound proceeding from a source with cones of silence and corresponding nodes or lines in which those cones cut the spherical wave surface. It is interesting to see that even in the neighborhood of the nodes the vibration is still perpendicular to the wave surface; so that we have realized in any case a gradual falling off of the intensity of the wave to zero and a passing through zero, which would be equivalent to a change of phase, without any motion perpendicular to the radius vector.

The more complicated terms that I have passed over are those that are sensible in the neighborhood of the source. Suppose, for instance that you have a bell vibrating. The sound slipping out and in over the sides of the bell and around the opening gives rise to a very complicated state of motion close to the bell; and similarly with respect to a tuning fork. If you take a spherical body, you may somewhat nearly express the motion in terms of spherical harmonics, and so on. You can see that in the neighborhood of the sounding body there will be a great deal of slipping in directions perpendicular to the radius vector, the displacements along the radius vector being compounded with motions out and in; but it is interesting to notice that all these motions become insensible at distances from the center large in comparison with the wave length. It is the consideration of these motions at distances that are moderate in comparison with the wave length that Stokes has made the basis of that very interesting investigation with reference to Leslie's experiment of a bell vibrating in a vacuum, to which I have already referred.

We may just notice, before I pass away from the subject, two or three points of the case, with reference to a tuning fork, a bell, and so on. Suppose the sounding body to be a circular bell. In that case clearly, if the bell be held with its lip horizontal, and if it be kept vibrating steadily in its gravest ordinary mode, the kind of vibration will be this: a vibration from a circular figure  into an elliptical figure  along one diameter, and a swinging back through the circular figure  into an elliptical figure  along the other diameter at right angles to the first. Clearly there would be practically a plane of silence here and another at right angles to it here (represented on the diagrams by dotted lines). Hence the solution for the radial component corresponding to this case, at a considerable distance from the bell, is $R = (\frac{1}{2} - \cos^2\theta)\frac{\cos\psi}{r}$, in order that the component may vanish when $\cos^2\theta = \frac{1}{2}$, or $\theta = \pm 45°$

On the other hand a tuning fork vibrating to and fro or an elongated elliptical bell (shaped like that which I have, which was obtained from that fine old Frenchman, Koenig's predecessor, Marloit, that makes an exceedingly loud sound), has an advantage in acoustic experiments over a circular bell. If you set a circular bell to vibrating and leave it to itself you always hear a beating sound, because the bell is not quite symmetrical. Excite it with a bow, and take your finger off, and leave it to itself, and if you do not choose the proper place to touch it, so that no vibration will occur there when you take your finger off, it will execute the resultant of two fundamental modes.

I do not know whether that experiment with plates is familiar to all of you. I would be glad to know whether it is. I make it always before my own classes, in illustrating the subject. Take a circular plate — just one of the ordinary circular plates that are prepared. Excite it in that way putting the finger on to make the quadrantal vibration .

That would be a case to which this solution obtained for the bell would apply. According to that notation the axis of x would be in this direction  X for a circular plate with two lines of silence at X right angles to one another. If sand be sprinkeled upon the plate, and I take my finger off, the sand at that point begins to oscillate and I hear a beating sound. But by a little trial, I find one place where if I touch the finger, and excite it so as to make a quadrantal vibration if I then take off my finger the sand remains undisturbed and there will be no beat. Then having found one place, I know there will be another place which is got by touching it here 180° from the first place, and that I can get another pair of nodes perpendicular to the first pair where there is also silence. Put your finger in between those places, force the plate to vibrate and take off your finger and you will have very loud beats, because the vibrations of the plate are not equal, — the two sounds always differ from one another. Try it in that way, and you will find it a most interesting thing with reference to circular plates. I have never seen it in any book, and I have done it every year for 45 years.

Take a division of the circumference into six equal parts by three diameters, and you find the same thing over again. Go on by trial touching the plate at two points 60° assunder, and bowing it 30° from either and you will get a sound resting on the three diameters determined by your fingers. Take off your fingers and you will in general get a beat. Follow your way around, little by little — it is very pretty when you come near a place of no beat. The moment you take off your fingers, you see the lines of nodes going backwards and forwards with a very slow oscillation. Get exactly the position, bow it, take off your finger, and the lines remain absolutely still. Take a point midway between those two and another 60° from that and you have a beat from loud sound to silence. If you try for it until you get exactly the intermediate position you will have the strongest beat possible, which is a beat from loud sound to silence. Advance your fingers another 30° and you will find the nodes remain absolutely still when you remove your fingers. You may go on in this way with eight and ten subdivisions, and so on; but you must not expect that the places for the octantal subdivisions, correspond to the places for the quadrantal subdivisions. The places for quadrantal subdivision will not in general be places for octantal subdivision. You must experiment separately for the octantal places and you will find generally that their diameters are oblique to the quadrantal.

The reason for all this is quite obvious. In each case, the plate being only approximately circular and symmetrical, the general equation for the motion has two approximately equal roots corresponding to the nodes or divisions by one, two, three, or four diameters, and so on. Those two roots always correspond to sounds differing a little from one another. The effect of putting the finger down at random is to cause the plate, as long as your finger is on it, to vibrate forcibly in a simple harmonic vibration of period greater than the one root and less than the other. But as soon as you take your finger off, it follows the law of superposition of fundamental modes; each fundamental mode being a simple harmonic vibration. I have often tried musicians with two notes which were very nearly equal, and said to them, "now, which of the two notes is the graver?" Sometimes they could tell, if the difference was not too small; but they are not accustomed to listen to the thing with physical ears, and do not always say which is the graver note. A person can tell at once, after having made a few experiments of that kind, that this is the graver and that the less grave note, even though he may have what musicians call an uncultivated ear, or a very bad ear for music, not good enough in fact to guide him in singing or make him sing in tune. It is very curious, when you have two notes which you thoroughly know are different, that if you sound first one and then the other, most people will say they are about the same. But sound them both together, and then you hear the discord of the two notes in approximate unison.

We need not go further into these divisions of disturbance in air. In every case there is a plane of silence. If you take a square plate or bell vibrating in a quadrantal mode, for instance then you have two vertical planes of silence at right angles to one another. If you make it vibrate with six or more subdivisions, you will have a corresponding number of planes of silence. I may go more into the case of the tuning fork. We have in general for quadrangular vibrations $R = (A - \cos^2\theta)\frac{\cos q}{r}$ where A is an unknown constant. In the particular case we have been considering, that constant is essentially $= \frac{1}{2}$.

With reference to the motions in the neighborhood of the tuning fork, you get this beautiful idea, that we have essentially harmonic functions to express them. Essentially algebraic functions of the coordinates appear in these distant terms, but in the other terms which Prof. Stokes has worked out, and which has been worked out in Prof. Rowland's paper on Electo-Magnetic disturbances* in a very full way, quite that kind of analysis appears, and it is ~~a~~ most important. I have not given you that part but only called your attention to the part with reference to the distance equation, partly because I think it is interesting for sound and partly because it prepares us for our special subject, waves of light.

---

* Phil. Mag, XVII, 1884, p. 413.
Am. Jour. Math. VI, 1884, p. 359

Tomorrow I think we shall begin and try to get sources ~~of waves~~ of waves of light. I want to lead you up to the idea of what the simplest element of light is. It must be polarized, and it must consist of a single sequence of vibrations. A body gets a shock so as to vibrate; that body of itself then constitutes the very simplest source of light that we can have; it produces an element of light. An element of light consists essentially in a sequence of vibrations. It is very easy to show that, and to prove that the velocity of propagation of sequences in the luminiferous ether is constant. It goes on, only varying with the variation of the source. As the force gradually subsides in giving out its energy, the amplitude evidently decreases; but there will be no shooting off of waves forward, no spreading out or lagging in the rear, no ambiguity as to the velocity. But when that comes into collision with other bodies, what is the result? According to the discussion to which I have referred, the velocity should be quite uncertain, depending upon the number of waves in the sequence, and all this seems to present a complicated problem.

But I am anticipating a little. We shall speak of this hereafter. Somebody asked me if I was going to get rid of the subject of groups of waves. I do not see how we can ever get rid of it in the wave theory of light. We must try to make the best of it, however.

This question of the vibration of particles is a peculiarly interesting and important problem. I hope you are not tired of it yet. You see that it is going to have many applications. In the first place it is at the base of the theory of the propagation of waves. When we take our particles uniformly distributed and connected by constant springs we may pass from the solution of the problems for the mutual influence of a group of particles to the theory, say, of the longitudinal vibrations of an elastic rod, or, by the same analysis, to the theory of the transverse vibrations of a cord.

I am going to refer you to Lagrange's Mecanique Analytique [Part 2. p. 339]. The problem that I put before you here is given in that work under the title of vibrations of a linear system of bodies. Lagrange applies what he calls the algorithm of finite differences to the solutions. The problem which I put before you is of a much more comprehensive kind; but it is of some little interest to know that cases of it may be found ramifying into each other.

I want to put before you some properties of the solution which are of very great importance. I want you to note first the number of terms.

We have: $c_j x_{j-1} = -a_j x_j$, $c_{j-1} x_{j-2} = -a_{j-1} x_{j-1} - c_j x_j = \frac{a_j a_{j-1}}{c_j} x_j - c_j x_j$, etc. All the x's being expressed in this way in terms of x_j, let N_i be the number of terms in x_{j-i}. These terms are obtained by substituting the values of x_{j-i+1}, x_{j-i+2} in the formula $-c_{j-i+1} x_{j-i} = a_{j-i+1} x_{j-i+1} + c_{j-i+2} x_{j-i+2}$. None of the terms can destroy one another except for special values, and the conclusion is that we have the following formula for obtaining the number of terms: $N_i = N_{i-1} + N_{i-2}$.

This is an equation of finite differences. Apply the algorithm of finite differences, as Lagrange says; or we may try for solutions of this equation by the following formula: $N_i = z N_{i-1}$. We thus find $z^2 = z + 1$, or $z = \frac{1 \pm \sqrt{5}}{2}$. We can satisfy our equation by taking either the upper or the lower sign. The general solution is, of course, $N_i = C\left(\frac{1+\sqrt{5}}{2}\right)^i + C'\left(\frac{1-\sqrt{5}}{2}\right)^i$ where C, C' are to be determined by the equation $N_0 = 1$, $N_1 = 1$. It is rather curious to see an expression of this kind for the number of terms in a determinant. You will find that the following is a solution of the more general equation $N_i = a N_{i-1} + b N_{i-2}$, viz:

$N_i = N_1 \frac{(r^i - s^i)}{r - s} + b N_0 \frac{(r^{i-1} - s^{i-1})}{r - s}$, where r, s are the two roots of the equation $x^2 = ax + b$. The coefficients of N_0, N_1, must of course be integral functions of $r + s = a$ and

$rs = -b$. If one of the roots be a proper fraction, s, for example, we may omit the large powers of s, and therefore for large values of i we may be sure of obtaining N_i to within a unit by calculating the integral part of $\frac{N_i r^i + N_0 b r^{i-1}}{r + \frac{b}{r}}$. The values of N_i up to $i = 12$ for the case of our problem ($a = b = N_0 = N_1 = 1$) are,

$i = 2, 3, 4, 5, 6, 7, 8, 9, 10, 11, 12.$

$N_i = 2, 3, 5, 8, 13, 21, 34, 55, 89, 144, 233.$

Lecture VII.

Lagrange, in the second section of the second part of his Mecanique Analytique on the Oscillation of a linear system of bodies, has worked out very fully the motion in the first place for disjoined bodies, and secondly for bodies forming a continuous cord. The case that we are working upon is not restricted to equal masses and equal connecting springs, but includes this particular linear system of Lagrange, in which the masses and springs are equal. I hope to take up that particular case, as it is of great interest. We shall take up this subject first to-day and the propagation of disturbances in an elastic solid second.

It was pointed out by Dr. Franklin that the formula for $N_i = aN_0 \frac{r^i - s^i}{r - s} + bN_0 \frac{r^{i-1} - s^{i-1}}{r - s}$ (which is equivalent to assuming $N_{-1} = 0$, so that $N_1 = aN_0$) may be thus simplified.

We have $r^2 = ar + b$, or multiplying by r^{i-1}, $r^{i+1} = ar^i + br^{i-1}$. So that the expression simplifies down to $N_i = N_0 \frac{r^{i+1} s^{i+1}}{r-s}$. *

We have, for example, $r - s = \sqrt{5}$, $r = \frac{1+\sqrt{5}}{2} = 1.618$. If we work this out by very moderate logarithms for the case $N_{12} = \frac{r^{13}}{r-s}$, dropping s^{13}, we find $13 \log 1.618 - \log \sqrt{5}$ $= 13 \times .209 - .3495 = 2.3675 = \log 233$, which comes out exact. This working with only 4 place logarithms.

I want to call your attention to something far more important than this. The dynamical problem, quite of itself, is very interesting and important, connected as it is with the whole theory of modes and sequences of vibration: but the application to the theory of light, for which we have taken this subject up gives to it more interest than we could have for it as a mere dynamical problem. I want to justify a fundamental form into which we can put our solution, which is of importance in connection with the application we wish to make.

Algebra shows that we must be able to throw $\frac{-x_1}{\xi}$ into the form

$$\frac{q_1}{\frac{\kappa^2}{\frac{1}{\tau^2}} - 1} + \frac{q_2}{\frac{\kappa_2^2}{\frac{1}{\tau^2}} - 1} + \dots \frac{q_j}{\frac{\kappa_j^2}{\frac{1}{\tau^2}} - 1}$$

where $q_1, q_2, \dots q_j$ are some constants, and $\kappa_1, \kappa_2, \dots \kappa_j$ are the values of the period τ for which $\frac{-x_1}{\xi}$ becomes infinite. We can put it into this form certainly, for if x_1, ξ be expressed in terms of x_j, they will be functions of the $(j-1)$st and jth degrees, respectively, in $\frac{1}{\tau^2}$. This is easily seen if we notice that $x_{j-1} = -\frac{a_j}{c_j} x_j$ is of the first degree in $\frac{1}{\tau^2}$, and that the degree of each x is raised a unit above that of the succeeding x by the factor $a_j = \frac{m_i}{\tau^2} - c_i - c_{i+1}$ in the equation $-c_i x_{i-1} = a_i x_i + c_{i+1} x_{i+1}$. Therefore, writing z for $\frac{1}{\tau^2}$, we have $\frac{-x_1}{c_1 \xi} = \frac{Az^{j-1} + A'z^{j-2} + \dots}{Bz^j + B'z^{j-1} + \dots}$, which on being expanded into partial fractions becomes $\frac{c_1}{\frac{1}{\tau^2} - \frac{1}{\kappa_1^2}} + \frac{c_2}{\frac{1}{\tau^2} - \frac{1}{\kappa_2^2}} + \dots \frac{c_j}{\frac{1}{\tau^2} - \frac{1}{\kappa_2^2}}$ which takes

* This may be obtained directly, by determining C, C', in terms of N_0 and $N_{-1} = 0$.

the required form on putting $C_i \kappa_i^2 = q_i$.

We know that the roots of the equation of i'th degree in $\mathcal{T}$ which makes $\frac{-x_1}{C_1\xi}$ becomes infinite are all real; they are the periods of vibration of a system of connected bodies. We have formal proof of it in the work which we have gone through in connection with such a system. I am putting our solution in this form, because it is convenient to look upon the characteristic feature of the ratio of $\mathcal{T}$ to one or other of the fundamental periods. In the first place it is obvious that if we know the roots $\kappa_1, \kappa_2, \ldots$ the determination of $q_1, q_2, \ldots$ is algebraic. Another form which I shall give you is an answer to that algebraic question, what are the values of $q_1, q_2 \ldots$ It is an answer in a form that is particularly appropriate for our considerations because it introduces the energy of the vibrations of the several fundamental modes in a remarkable manner. We will just get that form down distinctly.

Take the differential coefficients of $\frac{C_1\xi}{-x_1}$ with respect to $\frac{1}{\mathcal{T}^2}$, writing this form for the moment $\frac{q_1}{D_1} + \frac{q_2}{D_2} + \ldots$ Thus $\frac{d}{d\mathcal{T}^{-2}} \frac{C_1\xi}{-x_1} = \frac{\kappa_1^2 q_1/D_1^2 + \kappa_2^2 q_2/D_2^2 + \ldots}{(q_1/D_1 + q_2/D_2 + \ldots)^2}$; For the case $\mathcal{T} = \kappa_1$, our differential coefficient becomes $\frac{\kappa_1^2}{q_1}$, which determines $q_1 = \kappa_1^2 \Big/ \frac{d}{d\mathcal{T}^{-2}} \frac{C_1\xi}{-x_1}$. Now you will remember that we had $\frac{d}{d\mathcal{T}^{-2}} \frac{C_1\xi}{-x_1} = m_1 + m_2\left(\frac{x_2}{x_1}\right)^2 + \ldots m_j\left(\frac{x_j}{x_1}\right)^2$ For the moment, take the expression for the simple harmonic motion, and you see at once that that comes out in terms of the energy. Adopt the temporary notation of representing the maximum value by an accented letter. Then we have at any time of the motion $x_1 = x_1' \sin \frac{2\pi t}{\mathcal{T}}$, if we reckon our time from the time of each particle passing through its middle position, remembering that all the particles pass the middle position at the same instant. We have therefore for the velocity of particle No. 1, $\dot{x}_1 = \frac{2\pi}{\mathcal{T}} x_1' \cos \frac{2\pi t}{\mathcal{T}}$

The energy, which at any time is partly kinetic and partly potential, will be all kinetic at the moment of passing through the middle position. Take then the energy at that moment. For $t = 0$ we have $x_1 = 0$, $\dot{x}_1 = \frac{2\pi}{T} x_1'$. Denoting the whole energy by E (and remembering ~~that~~ that the mass $= \frac{m_1}{4\pi^2}$) we have

$$E = \frac{1}{2}\left(m_1 x_1'^2 + m_2 x_2'^2 + \dots m_j x_j'^2\right)\frac{1}{T^2}$$

Thus, the ratio of the whole energy to the energy of the first particle $\left(\frac{E}{\frac{1}{2}\frac{m_1 x_1'^2}{T^2}}\right)$ being denoted by R^{-1}, we have $m_1 R^{-1} = \frac{d}{dT^2}\frac{C_1\xi}{-x_1}$. This is true for any value of T whatever.

From this equation find, then, the ratios of the whole energy to the energy of the first particle when $T = \chi_1, \chi_2, \dots$. Denoting these several ratios by R_1^{-1}, R_2^{-1}..., we find $q_1 = \frac{\chi_1^2 R_1}{m_1}$, $q_2 = \frac{\chi_2^2 R_2}{m_1}$, ... Our solution becomes then,

$$\frac{-x_1}{C_1\xi} = \frac{T^2}{m_1}\left(\frac{\chi_1^2 R_1}{\chi_1^2 - T^2} + \frac{\chi_2^2 R_2}{\chi_2^2 - T^2} + \cdots\right)$$

This is the much more convenient form, as it shows us everything in terms of quantities whose determinations are suitable, viz: the periods, and energy ratios.

It remains, lastly, to show how, from our process without calculating the determinants, we can get everything that is here concerned. Our process of calculating gives us the u's in order, beginning with u_{j-1}. That gives us the x's in order, and thus we have all that is embraced in the differential coefficient with respect to $\frac{1}{T^2}$. Everything is done, if we can find the roots. I will show how you can find the roots from the continued fraction, without working out the determinant at all. The calculation in the neighborhood of a root gives us the train of x's corresponding to that root and then by multiplying the squares of the ratios of the x's to x_1 by the masses and adding, we have the corresponding energy.

The case that will interest us most will be the successive masses greater and greater, and the successive springs stronger and stronger, but not in proportion to the masses so that the periods of vibration of limited portions of the higher numbered particles of the linear system shall be very large. For example, so that if we hold at rest particles 4 and 6, the natural time of vibration of particle 5 will be longer than No. 2's would be if we held No's 1 and 3 at rest and set No. 2 to vibrating.

We will just put down once more two or three of our equations: $\frac{c_1 \xi}{-x_1} = a_1 - \frac{c_2^2}{u_2}, \ldots u_i = a_i - \frac{c_{i+1}^2}{u_{i+1}}$; $a_i = \frac{m_i}{\tau^2} - c_i - c_{i+1}$.

Without considering whether u_{i+1} is absolutely large or small, let us suppose that it is large in comparison with c_{i+1}. u_i will then be of the order a_i, u_{i-1} of the order a_{i-1}, and so on. We are to suppose that $a_1, a_2, \ldots a_i$ are in ascending order of magnitude. Now, $u_1 u_2 \ldots u_i = (-)^i c_1 \ldots c_i \frac{\xi}{x_i}$. We thus have this important proposition that the magnitudes of the vibrations of the successive particles decrease from particle No. 1 towards No. j; and x_i is exceedingly small in comparison with ξ, even though there is only a moderate proportion of smallness with respect to the ratios $\frac{c_1}{u_1}, \frac{c_2}{u_2}, \ldots \frac{c_i}{u_i}$. Thus see how small is the motion at a considerable distance from the point at which the excitation takes place, under the suppositions that we have been making.

Now, as to to the calculations. I do not suppose anybody is going to make these calculations; but I always feel in respect to arithmetic somewhat as Green has expressed in reference to analysis. I have no satisfaction in formulas unless I feel their arithmetical magnitude, – at all events, when formulas are intended for operations of that kind. So that if I do not exactly calculate the formulas, I would like to know how I would calculate them and express the order of the magnitudes. It might not be worth while to go into the number of terms *per se*, but the number of terms is closely related to the order of the magnitudes we have been dealing with. We are not going to make the calculations, but you will remark that we have every facility for doing so. In the first place, is the exceeding rapidity of convergence of the formulas. The question is to find $\frac{C_1\xi}{-x_1}$; everything, you will find, depends upon that. The exceeding rapidity of the convergence is manifest. Since u_2 is large, u_1 is equal to a_1 with a small correction; similarly $u_2 = a_2$ with a small correction and so on; so that two or three terms of the continued fraction will be sufficient for calculating the ratio denoted by u_1. The continued fractions converge with enormous rapidity upon the suppositions we have been making. We thus know the value of the differential coefficient $\frac{d u_1}{d \tau^2}$. We can in this way obtain several values of u_1 and begin to find it coming near to zero. Then take the usual process. Knowing the value of the differential coefficient allows you to diminish very much the number of trials that you must make for calculating a root. The process of finding the roots of this continued fraction will be quite analagous to Newton's process for finding the roots of an algebraic equation; and I tell any of you who may intend to work at it, that if you choose any particular case you will find that you will get at the roots very quickly.

I should think something like an arithmetical laboratory would be good in connection with class work in which students might be set at work upon problems of this kind, both for results, and in order to obtain facility in calculation. I think we will not say anything more about this problem just now, and we will leave it as we have it.

I hinted to you in the beginning about the kind of view that I wanted to take of Molecules connected with the luminiferous ether and affecting by their inertia its motions. I find since then that Lord Rayleigh really gave in a very distinct way, the first indication of the explanation of anomalous dispersion

I will just read a little of the paper on the Reflection & Refraction of Light by intensely Opaque Matter. [Philosophical Mag. May, 1872]. He commences, "It is, I believe, the common opinion, that a satisfactory mechanical theory of the reflection of light from metallic surfaces has been given by Cauchy, and that his formulae agree very well with observation. The result, however, of a recent examination of the subject has been to convince me that, at least in the case of vibrations performed in the plane of incidence, his theory is erroneous, and that the correspondence with fact claimed for it is illusory, and rests on the assumption of inadmissable values for the arbitrary constants. Cauchy, after his manner, never published any investigation of his formulae, but contented himself with a statement of the results and of the principles from which he started. The intermediate steps, however, have been given very concisely and with a command of analysis by Eisenlohr (Pogg. Ann. vol. CIV. p. 368), who has also endeavored to determine the constants by a comparison with measurements made by Jamin. I propose in the present communication to examine the theory of reflection from thick metallic plates, and then to make some remarks on the action on light of a thin metalic layer, a subject which has been treated experimentally by Quincke.

The peculiarity in the behavior of metals towards light is supposed by Cauchy to lie in their opacity, which has the effect of stopping a train of waves before they can proceed for more than a few wave-lengths within the medium. There can be little doubt that in this Cauchy was perfectly right; for it has been found that bodies which, like many of the dyes exercise a very intense selective absorption on light, reflect from their surfaces in excessive proportion just those rays to which they are most opaque. Permanganate of potash is a beautiful example of this given by Prof. Stokes.

He found (Phil. Mag. Vol VI, p. 293) that when the light reflected from a crystal at the polarizing angle is examined through a Nicol held so as to extinguish the rays polarized in the plane of incidence, the residual light is green, and that, when analyzed by the prism, it shows bright bands just where the absorption-spectrum shows dark ones. This very instructive experiment can be repeated with ease by using sunlight, and instead of a crystal a piece of ground glass sprinkled with a little of the powdered salt, which is then well rubbed in and burnished with a glass stopper or otherwise. It can without difficulty be so arranged that the two spectra are seen from the same slit one over the other, and compared with accuracy.

With regard to the chromatic variations it would have seemed most natural to suppose that the opacity may vary in an arbitrary manner with the wave-length, while the optical density (on which alone in ordinary cases the refraction depends) remains constant, or is subject only to the same sort of variations as occur in transparent media. But the aspect of the question has been materially changed by the observations of Christiansen and Kundt (Pogg. Ann. vols. cxli., cxliii, cxliv.) on anomalous dispersion in Fuchsin and other coloring-matters, which show that on either side of an absorption-band there is an abnormal change in the refrangibility (as determined by prismatic deviation) of such a kind that the refraction is <u>increased</u> below (that is, on the red side of) the band and <u>diminished</u> above it. An analogy may be traced here with the repulsion between two periods which frequently occurs in vibrating systems. The effect of a pendulum suspended from a body subject to horizontal vibration is to increase or diminish the virtual inertia of the mass according as the natural period of the pendulum is shorter or longer than that of its point of suspension. This may be expressed by saying that if the point of support tends to vibrate more rapidly than the pendulum, it is made to go faster still, and vice versâ" — I cannot understand the meaning of that sentence at all. There is a terrible difficulty with writers in abstruse subjects to make sentences that are intelligible. It is impossible to find out from the words what they mean; it is only from knowing the thing that you can do so — "Below the absorption-band the material vibration is naturally the higher, and hence the effect of the associated matter is to increase (abnormally) the vertical inertia of the aether, and therefore the refrangibility. On the other side the effect is the reverse." Then follows a note, "See Sellmeier, Pogg. Ann. vol cxliii p. 272". Thus Lord Rayleigh goes back to Sellmeier and I suppose he is the originator of all this. "It would be difficult to exaggerate the importance of these facts from the point of view of theoretical optics, but it lies beside the object of the present paper to go further into the question here."

There is the first clear statement that I have seen. Prof. Rowland has been kind enough to get these papers of Lord Rayleigh for me, with an immense deal of trouble. An interminable number of books have been brought to me, and in every one of them I have found something very important.

Sellmeier, Lord Rayleigh, Helmholtz and Lommel seems to be about the order. Lommel does not quote Helmholtz. I am rather surprised at this, because Lommel comes three or four years after Helmholtz, 1874, and 1878 are the respective dates. Lommel's paper is published in Helmholtz's Journal [Ann. der Physik und Chemie 1878, vol 3, p. 339] so I suppose Helmholtz has no objection. Helmholtz's paper is excellent. Lommel goes into it still further, and has worked out the vibrations of associated matter to explain ordinary dispersion.

I only found this forenoon that Lommel [Ann. der Ph. und Chem. 1878, Vol. 4, p. 55] also goes on to double refraction of light in crystals – the very problem I am breaking my head against. He is satisfied with his solution, but I do not think it at all satisfactory. It is the kind of thing that I have seen for a long time, but could not see that it was satisfactory; and I do see reason for its not being satisfactory. He goes on from that and obtains an equation which would approximately give Huygens surface. I have not had time to determine how far it may be correct. The exceedingly close agreement of Huygen's surface with the facts of the case which Stokes has found absolutely cuts the ground from under a large number of ver tempting modes of explaining double refraction.

Lecture VIII

We shall take some fundamental solutions for wave motion such as we have already had considerable to do with, ~~only~~, only we shall consider them as now applicable to distortional waves, instead of condensational waves. That is, we can take our primary solution in the form $\varphi = \frac{1}{r} \sin \frac{2\pi}{\lambda}(r - ct)$, where $c = \sqrt{\frac{k + \frac{4}{3}n}{\rho}}$ if the wave is is condensational, and $= \sqrt{\frac{n}{\rho}}$ if the wave is distortional. But for a distortional wave, we must also have what is denoted by $\delta = 0$.

In the first place, we know that Φ satisfies $\rho \frac{d^2\Phi}{dt^2} = n \nabla^2 \Phi$, our value of c being $\sqrt{\frac{n}{\rho}}$ — (I want very much a name for that function ∇, — delta turned upside down. I do not know whether Prof. Ball has any name for it or not; Sir Wm. Hamilton uses it a great deal, and I think perhaps, Prof. Ball may know of a name for it). The conditions to be fulfilled by the three components of displacement, ξ, η, ζ, of a distortional wave are, in the first place, $\rho \frac{d^2\xi}{dt^2} = n \nabla^2 \xi$, $\rho \frac{d^2\eta}{dt^2} = n \nabla^2 \eta$, $\rho \frac{d^2\zeta}{dt^2} = n \nabla^2 \zeta$; and we must have besides $\frac{d\xi}{dx} + \frac{d\eta}{dy} + \frac{d\zeta}{dz} = 0$. Thus ξ, η, ζ, must be three functions, each fulfilling the same equation. There is a fulfilment of this equation by the functions Φ; and as we have one solution, we can derive other solutions from that by differentiation. Let us see then, if we can derive three solutions from this value of Φ which shall fulfil the remaining condition. It is not my purpose here to go into an analytical investigation of solutions; it is rather to show solutions which are of fundamental interest. Without further preface then, I will show you one, and another, and then I will interpret them both.

Take for example the following, which obviously fulfils the equation $\frac{d\xi}{dx} + \frac{d\eta}{dy} + \frac{d\zeta}{dz} = 0$:

$$\xi = 0, \quad \eta = -\frac{d\Phi}{dz}, \quad \zeta = \frac{d\Phi}{dy}.$$

In each case the distance terms only of our solution are what we wish. Thus

$$\eta = -\frac{d\Phi}{dz} \doteq -\frac{2\pi}{\lambda} \cdot \frac{z}{r^2} \cdot \cos q, \quad \zeta = \frac{d\Phi}{dy} \doteq \frac{2\pi}{\lambda} \cdot \frac{y}{r^2} \cdot \cos q.$$

Remark that in this solution the displacement at a distance from the source is perpendicular to the radius vector; i.e., we have $x\xi + y\eta + z\zeta = -y\frac{d\Phi}{dz} + z\frac{d\Phi}{dy} \doteq 0$. Before going further, it will be convenient to get the rotations. It is an exceedingly convenient way of finding the direction of vibration in distortional displacements. The rotations about the axes of x, y, z, will be:

$$\frac{d\zeta}{dx} - \frac{d\eta}{dy} \doteq -\frac{4\pi^2}{\lambda^2} \frac{y^2+z^2}{r^3} \sin q; \quad \frac{d\xi}{dz} - \frac{d\zeta}{dx} \doteq \frac{4\pi^2}{\lambda^2} \frac{xy}{r^3} \sin q; \quad \frac{d\eta}{dx} - \frac{d\xi}{dy} \doteq \frac{4\pi^2}{\lambda^2} \frac{xz}{r^3} \sin q.$$

These rotations are proportional to $\frac{x^2}{r^3} - \frac{1}{r}$, $\frac{xy}{r^3}$, $\frac{xz}{r^3}$; that is to say, besides the x component $-\frac{1}{r}$, we have an r component $\frac{x}{r^2}$. We have a rotation around the radius vector r, and a rotation around the axis of x, whose magnitudes are proportional to $\frac{x}{r^2}$ and $\frac{1}{r}$.

If you think out the nature of the the thing, you will see that it is this: a globe, or a small body at the origin, set to oscillating about Ox as an axis. You will have turning vibrations everywhere; and the light will be everywhere polarized in planes through Ox. The vibrations will be everywhere perpendicular to the radial plane through Ox.

In the first place we have (omitting the constant factor $\frac{2\pi}{\lambda}$) $\xi = 0$, $\eta = -\frac{z}{r^2}\cos q$, $\zeta = \frac{y}{r^2}\cos q$. That presents a wave spreading out in all directions from the axis of x. For if $y=0$, $z=0$, the displacements are zero, or we have the case of zero vibration in the axis of x. Again, the displacements are everywhere perpendicular to Ox (since we always have $\xi = 0$), and being perpendicular also to the radius vector, they are perpendicular to the radial plane through the axis of x.

Let us consider the state of things in the planes yz. Suppose we have a small body here at the origin, or center of disturbance, 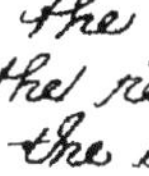 and that it is made to turn in this way (indicating a twisting motion about an axis perpendicular to the plane of the paper) in a given period T. What is the result? Waves will proceed out in all directions and the intersections of the wave front with the plane (yz) of the paper will be circles. We shall have vibrations perpendicular to the radius vector of magnitude $\frac{\cos q}{r}$, which is the same in all directions. The rotation, which is simply the polar rotation about the axis of x in the plane yz, is $\frac{2\pi}{\lambda}\,\frac{\sin q}{r}$ (also the same in all directions). There is therefore a minimum displacement where there is a maximum distortion, and *vice versa*. At a point of maximum distortion (positive or negative) there is zero rotation; at a point of maximum rotation there is zero distortion. We have polarized light consisting of vibrations in the plane and perpendicular to the radius vector, and therefore the plane of polarization is the radial plane through OX.

Here we have a simple source of polarized light; it is the simplest form of polarization and the simplest source that we can have. Every possible light consists of sequences of light from simple sources. Is it probable that the shocks to which the particles are subjected in the electric light, or in fire, or in any ordinary source of light would give rise to a sequence of this kind? No; because there is nothing to make a body vibrate by itself. We can arbitrarily do it, for we can do what we will with the particle. That privilege occurred to me in Philadelphia last week, and I showed the vibrations by having a large bowl of jelly made with a ball placed in the middle of it. I really think you will find it interesting enough to try it for yourselves. It allows you to see the vibrations we are speaking of. I wish I had it to show you just now so that you might see the thing in force. It saves brain very much.

I had a large glass bowl quite filled with some spirit jelly, and a wooden ball floating in the middle of it.

Try it and you will find it a very pretty illustration. Apply your hand to the ball, and give it a twisting motion thus, and you have exactly the kind of motion here expressed in the plane yz. The motion in any oblique direction such as at this point $(x\,y\,z)$ you will find to consist of polarized light vibrating perpendicular to the radial section. The amplitude of the vibration here (in the vertical axis) is zero; here at the surface (in the plane yz) it is $\frac{1}{r}\cos q$; and if you use polar coördinates, calling this angle θ (indicating on the diagram) then the amplitude here (at $x\,y\,z$) is $\frac{1}{r}\cos q \cdot \sin\theta$, giving, when θ is a right angle the previous expression.

I say that this is the simplest source and the simplest strain of polarized light that we can imagine. But it cannot be induced naturally, because no natural vibrator could do it. The next simplest is a globe or small body vibrating to and fro in one line. We will take the solutions for that presently. Still we have not got up to the essential complexity of the natural vibrator. I may take my hand and give torsional oscillations to the globe; I can take my hand (and that makes a very pretty modification of the experiment) and shove out on the globe, making it vibrate; and people cannot help saying, "O, there is the natural time of the vibrations, you find it if you only leave it alone to itself." But it is only proper for an illustration of vibrations spreading out from a center. We are bothered here also by reflections back, as it were, from the containing bowl; just as in suspending a rope to show waves running along it, we are bothered in the experiment by the rope not being infinitely long. You can always see a set of vibrations running along the rope, beginning at the lower end and reflected back from the upper end where it is fastened to the ceiling. But in this experiment, you do not see the waves travelling out at all because you get it in a certain set of vibrations, depending on this finite material. But just imagine the bowl to be infinitely large, and that you commence making torsional oscillations; what will take place? A spreading outwards of this kind of vibrations, the beginning being, as we shall see, abrupt. We shall scarcely reach that to-day, but we shall consider the abruptness of the beginnings and endings of the vibrations in an elastic solid; and in every case in which the velocity of propagation is independent of the wave length we have no end at all, but waves travelling outwards, with a gradual falling off of intensity.

When you apply your hands and force the ball to perform those torsional vibrations, you have waves proceeding from it; but if you then leave it to itself there is no vibrating energy in it at all, except the slight angular velocity that you leave it with. A vibrator which can send out a succession of impulses independently of being forced to vibrate from without, must be a vibrator with the means of conversion of potential into kinetic energy in itself. A tuning fork, and a bell, are sample vibrators in sound. The simplest sample vibrator that we can get to represent the origin of the simplest sequence of light is just like a tuning fork. Two bodies joined by a spring would be more symmetrical than a tuning fork. Two globes joined by spring - that will give you the idea; or (which will be a vibration of the same type still) one spherical body vibrating backwards and forwards from having been drawn so,  into an oval shape, and let go.

I will look, immediately, at a set of vibrations produced in an elastic solid by a sample vibrator. But suppose you produce vibrations in your jelly solid by taking hold of this ball and shoving it to and fro horizontally, or again shoving it up and down vertically, and think of the kinds of vibrations it will make all around. Think of that, in connection with the formulae, and it will help us to interpret them. But it will take a higher order of vibrator to get the kind of vibration that comes from the natural source. We might have those torsional vibrations; but among all the possible vibrations of atoms in the clang and clash of atoms that there is in a flame, or other source of light, a not very rare case I think would be that which I am going to speak of now. It consists of opposite torsional vibrations at the two ends of an elongated mass; or, to simplify our conception for a moment, imagine two globes connected by a columnar spring; twist them in opposite directions, and let them go. There might be a source of vibrations, and if the potential energy of the spring is very large in comparison with the energy that has been carried off in a thousand or a hundred thousand vibrations, you will have a set of vibrations following the same law that we get in the case already considered.

Before passing on to the to and fro vibrator we will think of this motion for a moment but we will not work it out, because it is not so interesting. To suit our drawing

we shall suppose one globe here, and another upon the opposite side on a level with the first, so that the line of the two is perpendicular to the board. Give these globes opposite torsional vibrations about their common axis, and what will the result be? A single one produces zero light in the axis and maximum light in the equatorial plane. The two going in opposite directions will produce zero light in the equatorial plane and zero light in the axis; so that you will proceed from zero in the equatorial plane to a maximum between the equatorial plane and the poles and zero at the poles, and you will have opposite vibrations in each hemisphere. That constitutes a possible case of vibrations of polarized light, proceeding from a possible independent vibrator. If you had, among all the elements concerned in the production of the light, some such action, or configuration as that if a shock took place at one end of a molecule, another should simultaneously take place in an opposite direction upon the other end, that might set the thing to vibrating in that way; and that is one of the possible sets of vibrations constituting light.

But by far the most simple and natural supposition in respect to an independent vibrator is afforded by the illustration of a bell, or a tuning fork, or an elastic body deformed from its natural shape and left to vibrate. In all these cases, you remark, the center of gravity of the vibrator is at rest; and you can not have anything else from an independent action. The vibrator must have potential energy in itself, and its center of gravity must be at rest except insofar as the reaction of the medium upon it causes a slight motion of the center of gravity.

I will put down the solution which corresponds to a to and fro vibration in the axis of x, viz.

$$\xi = \frac{4\pi^2}{\lambda^2}\phi + \frac{d^2\phi}{dx^2}, \quad \eta = \frac{d^2\phi}{dy\,dx} \quad \zeta = \frac{d^2\phi}{dz\,dx}.$$

ϕ is our old friend, $\frac{1}{r}\sin\frac{2\pi}{\lambda}\left(r - t\sqrt{\frac{n}{\rho}}\right)$ In the first place we know that $\rho\frac{d^2\xi}{dx^2} = n\nabla^2\xi$, etc, are satisfied, because ϕ and all its differential coefficients satisfy this relation. We have then only to verify that the dilitation is zero. I will not go through the verification, but you will not make the solution your own unless you see how I obtained it. I will not say that there is anything novel in it, but it is simply the way it occurred to me. I obtained it to illustrate Stokes's explanation of the blue sky. I afterwards found that Lord Rayleigh had gone into the subject more searchingly than Stokes, and I read his work upon it.

The way I found this solution was this: $\frac{d\phi}{dx}$ is clearly the displacement potential corresponding to a source of the kind, a pull along the axis of x. It is like the magnetic potential of a bar magnet with its axis in the direction of OX. The displacement function of which the displacements are the differential coefficients would take that form if this was a question, for instance, of sound and not of light. It was a question of condensational vibrations with us several days ago. I did not go into the matter in detail, but we saw that for condensational vibrations proceeding from a vibrator vibrating to and fro along the axis of x that $\frac{d\varphi}{dx}$ was the displacement potential and it is obvious, if we start from the very root of the matter that it must be so $\frac{d\varphi}{dx}$ must therefore be the corresponding function that we shall have to deal with in the case of light from such a source although that will not certainly give by differentiation simply, the displacements we want. The displacements in the condensational wave problem are displacements which fulfil certain of the conditions, but do not fulfil all the conditions, of giving us a pure distortional wave unless we add a term or terms in order to make the dilatation zero. Just try in the first place for the dilatation. We have $\nabla^2\varphi = \frac{\rho}{n}\frac{d^2\varphi}{dt^2} = -\frac{4\pi^2}{\lambda^2}\varphi$, in which we may substitute $\frac{d\varphi}{dx}$ for φ. Thus $\nabla^2\frac{d\varphi}{dx} = -\frac{4\pi^2}{\lambda^2}\cdot\frac{d\varphi}{dx}$. We have verified therefore, that the displacements given satisfy $\frac{d\xi}{dx}+\frac{d\eta}{dy}+\frac{d\zeta}{dz}=0$; and thus we have made up a solution which satisfies the condition of being non-condensational — no condensation or rarefaction anywhere.

In the first place, taking the distant terms only we have $\xi = \frac{4\pi^2}{\lambda^2}\frac{r^2-x^2}{r^3}\sin q$, $\eta = -\frac{4\pi^2}{\lambda^2}\frac{xy}{r^3}\sin q$, $\zeta = -\frac{4\pi^2}{\lambda^2}\frac{xz}{r^3}\sin q$. It is easy to verify that these displacements are perpendicular to the radius vector, i. e. that we have $x\xi+y\eta+z\zeta=0$. Just look at the case along the axis of x, and again in the plane yz. It is written down here in mathematical word painting as clearly and completely as any non-mathematical words can give it. Take $y=0$, $z=0$, and that makes $\xi=0$, $\eta=0$ $\zeta=0$. Therefore, in the direction of the axis of x there is no motion. That is a little startling at first, but is quite obviously a necessity of the fundamental supposition. Cause a globe in an elastic solid to vibrate to and fro. At ~~the very~~ surface of the globe the points in which it is cut by OX have the maximum motion; and through out the whole circumference of the globe; the medium is pulled, by hypothesis, along with the globe. But this is not a solution for that comparatively very difficult problem. I am only asking you to think of this as the solution for the motion at a great distance. It may not be a globe, but a body of any shape moved to and fro. To think of a globe will be more symmetrical. In the immediate neighborhood of the vibrator there is a motion produced in the line of vibration; the motion of the elastic solid in that neighborhood consists in a somewhat complex, but very easily expressed state of things, in which we have particles in one place, moving out and slipping around with motions oblique to the radius vector, as in the axis of x, and in other places moving perpendicular to the radius vector, as for points in the plane yz. All however, except motions perpendicular to the radius vector, become insensible at distances very great in comparison with the wave length. We have taken, simply, the leading terms of the solution. These represent the motion at great distances, quite irrespective of the shape of the body, and the comparatively complicated motion in the neighborhood of the vibrating body.

Take now $x=0$, and think of the motions in the plane yz. The vibrator is supposed to be vibrating perpendicular to this plane. We have $\xi \doteq \frac{4\pi^2}{\lambda^2}\frac{1}{r}\sin q$, $\eta = 0$, $\zeta = 0$. What does that mean? Clearly, that the vibrations are perpendicular to this plane yz. We have light spreading out uniformly in that plane, and polarized in that plane, the vibrations being perpendicular to it. That is what Stokes supposed was necessarily the theory of the blue light of the sky. Lord Rayleigh showed that it was not so clear as Stokes had supposed. He elaborately investigated the question "Is the blue light of the sky — which was supposed to be owing to particles which you may say are spherical if you like — due to the particles causing it being of density different from the surrounding luminiferous ether, or being of rigidity different from the surrounding luminiferous ether." The real question would be, If the particles are water, what is the theory of waves of light in water; does it differ from air in being, as it were, a denser medium with the same effective rigidity, or is it a medium of the same density and less effective rigidity, or will both density and rigidity vary?

Lord Rayleigh examined that question very thoroughly and finds, if the cause were, for instance, little spherules of water and if in the passage of light through water the fact that the velocity of propagation is slower than in air were explained by less rigidity and the same density we should have something quite different in the polarization of the sky from what we would have on the other supposition. On the other hand, the polarization of the sky creates the supposition (which is as much as the uncertaintitude of the experimental data allows us to judge) that the particles, whether they be particles of water, or motes of dust or whatever they may be, act as if they were little portions of the luminiferous ether of greater density and not of rigidity, than the surrounding ether.

This solution, then, is not the solution for that source of light which has such great interest as being the cause of the blue light coming from the sky. I will call attention a little more to Lord Rayleigh's explanations upon that; but it cannot be the effect of a vibrator in the source, for the reasons I have stated. We may differentiate once more with respect to x, in order to get a proper form of function that will express the motion from the vibrator vibrating to and fro like this — vibrating the hands towards and from each other. Then we shall have a vibrator which will express one single sequence of vibrations, of which the multitude constitutes the light of the source. The question is then forced upon us, what is the velocity of a group of waves in the luminiferous ether undisturbed by ordinary matter. With a constant velocity of propagation each group remains unchanged. But how about the effect of a non-simple source of light in a transparent medium like glass? It is a question that is more easily put than answered. We should consider it carefully. I do not dispair of seeing the answer. I think, if we have a little more patience with our dynamical problem we shall get it.

Here is a perfectly parallel problem. Commence suddenly to give a simple harmonic motion through the handle P to our system of particles $m_1, m_2, \ldots m_j$, which play the part of a molecule, of course. If you commence suddenly imparting to the handle a motion of any period whatever, avoiding only one of the fundamental periods, if there be a little viscosity it will settle into a state of things in which you have perfectly regular simple harmonic vibration. If there be no viscosity whatever, what will the result be? It will be the component of simple harmonic motions in the period of our applied motion at the bell handle P; with every part in it obtained by a continued fraction. We superimpose motion upon it, and jangle it as it were, producing coexistent simple harmonic vibrations of the fundamental periods. If there is no viscosity, that state of things will go on forever. I cannot satisfy myself with viscous terms in these theories (although I believe this is the view of Lommel, Helmholtz and others) because we know that light goes on for millions and millions of vibrations. But if we have none of these viscous terms at all whatever velocity we have must show in the vibration of something else, and that is what? In going into that sort of vibration with which we have been occupied in the other part of our course, we must account for

~~it~~ these irregular vibrations somehow or other. The viscous terms are merely a step towards accounting for the difficulties of the theory. By viscous terms, I mean terms that assume a viscosity.

But the state of things with us is that that jangling will go on forever, if there is no loss of energy; and we want to coax our system of vibrators into a state of vibration with an arbitrarily chosen period without viscous consumption of energy. Begin thus: get it into motion with a very small range. The result will be just as I have said, only with a very small range. After waiting a little time increase the range; after waiting a little longer, increase the range again, and so go on, increasing the range by successive steps. Each of those will superimpose another state of vibration. There would be, I believe, virtually an addition of the energies of those several vibrations if you make these steps quite independent of one another.

For example, suppose you proceed thus. In the first place, start right off into vibrations of your handle P through a space, say of 30 inches. You will have a certain amount of energy in the irregular vibrations. In the second place, commence on a range of three inches. After you have kept it going on three inches any time you like, suddenly increase it to three inches more, making it six inches. Then, sometime after, suddenly increase the range to nine inches; and go on in that way for ten steps. The energy of the irregular vibrations produced by suddenly commencing through the range of three inches, which is one-tenth of 30 inches, will be one hundredth of the energy which you would have if you commenced right away with the vibration through 30 inches. Each successive step of three inches will add the one-hundredth; and the result is that if you go by these steps to the range of 30 inches, you will have in the irregular vibrations one-tenth of the energy you would get if you began at that range right away. Thus, by very gradually increasing the range, the result will be that there will be infinitely little of the irregular vibrations.

I believe something of that kind will account for our difficulty; and I believe that that kind of thing applied to sequences of waves will without doubt show that if you commence a set of waves very gradually, through several hundred vibrations may be enough, and then make them uniform (that is let the source go on uniformly after that) that even with sea waves possibly, or with luminous waves in a transparent solid, there will be exceedingly little disturbance from the beginning and end of them. It is only a vague idea I have thrown out; but I think considerations of this kind may help us to see how it is that sets of groups of waves which undoubtedly constitute the reality of light, do still act as if we had a perfectly simple harmonic and continuous source of vibrations. They do act so in the propagation of light through the medium, in refraction, and reflection, and so on.

But there are cases in which we have that tremendous jangling, and that is in the fluorescence of such a thing as uranium glass which lasts for several seconds after the exciting light is taken away, and then again in phosphorescence that lasts for hours and days. There have been exceedingly interesting beginnings, in the way of experiments already made, but I think nobody has found whether initial refraction is exactly the same as permanent refraction. For this purpose we might use Becquerel's phosphoroscope or we might take such an appliance as Prof. Michelson has been using for light and get something more enormously searching than Becquerel's phosphoroscope, and try whether in the first hundredth of a second there is any indication of a different wave velocity from that which you would have when light passes continuously in the usual manner of refraction. If in the methods employed for ascertaining the velocity of light in a transparent body, (notwithstanding the criticisms that they have received at the British Association meeting, to which I have referred several times) we apply a test for an instantaneous refraction, I have no doubt we shall not get negative results, but get properties of ultimate importance. We might take bodies in which, like uranium glass, the phosphorescence lasts only a few seconds. and then again bodies in which phosphorescence lasts for minutes and hours. With some of those we should have anomalous dispersion, gradually fading away after a time. I should think that by experimenting, and so on, we should find some very interesting results of this kind.

Lecture IX.

We shall go on for the present with the subject of the propagation of waves from a center. Let us pass to the case of two bodies vibrating in opposite directions, in the manner which we have already had for one body which was expressed by $\xi = \frac{4\pi^2}{\lambda^2}\varphi + \frac{d}{dx}\frac{d\varphi}{dx}$, $\eta = \frac{d}{dy}\frac{d\varphi}{dx}$, $\zeta = \frac{d}{dz}\frac{d\varphi}{dx}$. We verified that $\frac{d\xi}{dx} + \frac{d\eta}{dy} + \frac{d\zeta}{dz} = 0$ so that this expresses rigorously a distortional wave. It is obvious that this expresses the result of a two and fro motion at the origin. Remark, for one thing, that in the neighborhood of the origin, at such moderate distance from it that the component motion in the direction OX does not vanish, we have on the two sides of the origin simultaneously positive values. ξ is the same for a positive value of x as for x the negative of that value. At distances from the origin in the line OX which are considerable in comparison with the wave length the motion vanishes as we have seen. This, then, expresses the result of a to and fro motion at the origin.

Pass on now, to this case: a positive to and fro motion on the one side of the origin, and a negative to and fro motion on the other side of the origin. I will indicate these motions by arrow heads. The first case already considered 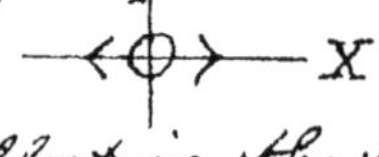

; the second case 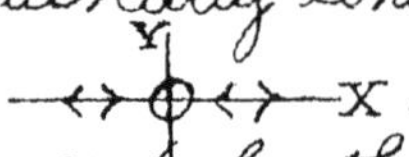

. The effect in the first case being expressed by the displacements ξ, η, ζ, already given, the effect in the second case will be expressed by the displacement $\frac{d\xi}{dx}, \frac{d\eta}{dy}, \frac{d\zeta}{dz}$. This displacement clearly expresses a motion which has opposite signs for equal positive and negative values of x. It will express a simultaneous outward and inward motion on the two sides of the origin and a zero motion in the plane yz. A motion for distances from the origin moderate in comparison with the wave length, will be accurately expressed by these functions; but as before we shall take only the leading or distance terms, and also ~~the~~ drop the coefficient $-\frac{8\pi^3}{\lambda^3}$ which we do not want. Thus $\xi = \frac{(x^2 - r^2)x}{r^4}\cos q$, $\eta = \frac{x^2 y}{r^4}\cos q$, $\zeta = \frac{x^2 z}{r^4}\cos q$ express the distant displacement of an outward and inward motion illustrated by that configuration of arrow heads (case 2), and obviously expressing a motion in which there will be zero displacement everywhere in the plane yz, with equal opposite values on the two sides of that plane. Take $y = 0$, $z = 0$, to find the motion in the axis of x, and we get as in the first case zero motion in that axis. We can easily satisfy ourselves that the radial component of the displacement is zero i.e. that we have $x\xi + y\eta + z\zeta = 0$. Lastly, if you think of the kind of polarization that will be produced by that motion, it is obvious that the motion will be everywhere symmetrical around the axis of x, and will be in the radial plane through OX. $[0.\xi + z\eta - y\zeta = 0]$ Therefore, we have light polarized in the plane through the radius of the point considered and perpendicular to the radial plane through OX

Look at what the magnitude of the motion will be

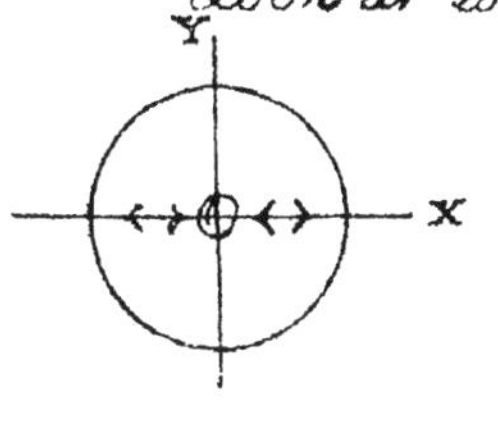

Inasmuch as the motion is symmetrical around the axis of x, we may take what goes on in the plane xy as a sample of the whole. We then have $\xi = -\frac{xy^2}{r^4}\cos q$, $\eta = \frac{x^2 y}{r^4}\cos q$; showing that there is zero motion in the axis of x, and zero motion in the axis of y. The expression for the amplitude of the motion is $\sqrt{\xi^2 + \eta^2} = \frac{xy}{t^3} \cos q$. Thus the displacement is distributed on the two sides of OX and of OY so as to be equal and opposite in adjacent quadrants. Remember that the thing is symmetrical around OX, and you have a perfect understanding of the distribution of the motion, the distance being considerable in comparison with the wave length.

This is the simplest set of vibrations that we can consider as proceeding from any natural source of light. As I said, we might conceive of a pair of equal and oppositely torsional motions, at the two ends of a vibrating molecule. That is one of the possibilities, and it would be rash to say that any one possible kind of motion does not exist in so remarkably complex a thing as the motion of the particles from which light originates.

This motion we are considering is perhaps the most interesting as it is obviously the simplest kind of motion that can proceed from a single vibrator. If you consider the two ends of a tuning fork, neglecting the prongs, so that everything may be symmetrical around the two moving bodies, you have a way by which the motion may be produced. Or our source might be two balls connected by a spring and pulled assunder and set to vibrating in and out; or it might be an elastic sphere which has experienced a shock. An infinite number of modes of vibration are generated when an elastic ball is struck a blow, but the gravest mode is also no doubt where the energy is greatest, and that consists of the globe vibrating from an oblate to a prolate figure of revolution.

The kind of thing that the luminiferous vibrator consists in seems to me to be a sudden initiation of a set of vibrations and a sequence of vibrations from that initiation which will naturally become of smaller and smaller amplitude. So that the graphic representation of what we should see if we could see what proceeds from one element of the source, the very simplest conceivable element of the source, would consist of polarized waves of light spreading out in all directions according to some such law as we have here. In any one direction, what will it be? Suppose that the wave advances from left to right; you will then see what is here represented on a magnified scale.

I have tried to represent a sudden start and a gradual falling off of intensity. Why a sudden start? Because I believe that the light of the natural flame or of the arc light, or of any other known source of light must be the result of sudden shocks from a number of vibrators. Take the light obtained by striking two quartz pebbles together. You have all seen that. There is one of the very simplest sources of light. There is some sort of a chemical or ozoniferous effect connected with it which makes a smell. As to what the cause of that may be, I suppose we are almost assured, now, that it proceeds from the generation of ozone. What sort of a thing can the light be that proceeds from striking two quartz pebbles together? Under what circumstances can we conceive a group of waves of light to begin gradually and to end gradually? You know what takes place in the excitation of a fiddle string or a tuning fork by a bow. The vibrations gradually set up from zero to a maximum and then, when you take the bow off, gradually subside. I cannot see anything like that in the source of light. On the contrary, it seems to me to be all shocks, a sudden beginning and gradual subsidence.

I say this, because I have just been reading a very interesting paper by Lommel, I think, or Sellmeier* (both touch upon this) which goes into the thing very fully. Helmholtz remarks that he gets into a little difficulty in his dynamics and does not show clearly what becomes of the energy in a certain case, but he takes hold of the thing with great power. He goes into this case very fully, and in the way with which we are all familiar. He remarks that Fizeau obtained a suite of 50,000 vibrations interfering with one another, and judges from that that ordinary light consists of polarized light, circular or elliptical, or plane polarized as I said to you myself, one or two days ago, with (what I did not say) the plane of polarization, or one or both axes of the ellipse if it be elliptically polarized, gradually changing, and the amplitude gradually changes. He says gradually and so gradually that there is not so great a change in the course of 50,000, or 100,000, or perhaps several million vibrations in the amplitude or mode of polarization as to prevent interference. In fact, I suppose there is no perceptible difference between the perfectness of the annulments with 50,000 vibrations than with 1,000; although I speak here not with confidence and I may be corrected. You have seen that, have you not Prof Rowland?

Prof. Rowland: Yes; but it is very difficult to get the interferences.

Sir Wm. Thomson: But when you do get them, the black lines are very black, are they not?

Prof. Rowland: I do not know. They are so very faint that you can hardly see them.

Sir Wm. Thomson: What do you infer from them?

Prof. Rowland: That there is a large number. The width of the lines of the spectrum indicates how perfectly the light interferes; and with a grating of very fine lines I find exceedingly perfect interference for at least 100,000 periods.

*Sellmeier; Ann. der Phy. u. Chem. 1872, vols. 145, 147.

I should think.

Sir Wm. Thomson: That goes further than Fizeau. Sellmeier says that probably a great many times 50,000 waves must pass before there can be any great change. He goes at the thing very admirably for the foundation of his dynamical explanation of absorption and anomalous reflection. The only thing that I do not fully agree with him in his fundamentals is the gradualness of the initiation of light at the source. I believe, in the majority of cases at all events, in sudden beginnings, and gradual endings. Prof. Rowland has just told us how gradual the endings are. Fizeau could infer that the amplitude does not fall off greatly in 50,000 vibrations. It is quite possible from all we know, that the amplitude may fall off considerably in 100,000 vibrations, is it not?

Prof. Rowland: The lines are then very sharp.

Sir Wm. Thomson. It would not depend on the sharpness of the lines, would it?

Prof. Rowland: O, yes. It would draw them out of line.

Sir Wm. Thomson: Would it broaden them out, or throw a little light over a place that should be dark?

Prof. Rowland: It would broaden them out.

Sir Wm. Thomson: It is a very interesting subject, and from the things that have been done by Prof. Rowland and others, we may hope to see if we live a knowledge of the difficulties quite incomprehensibly superior to what we have now. I doubt, however, whether we will live to see knowledge that we can have hardly any conception of now in the way of the extinction of vibrations in reference to light. We are perfectly certain that the diminution of amplitude must be exceedingly small — practically nil in 1000 vibrations; we can say that it is practically nil in 50,000 vibrations, we know that it is nearly nil in

100,000 vibrations. Is it practically nil in two or three hundred thousand vibrations, or in several million vibrations? Possibly not. Dynamical considerations come into play here. We shall be able to get a little insight into these things by forming some sort of an idea of the total amount of energy there must be in one vibrator, and what sequences of waves it can supply.

In speaking of Sellmeier's work and Helmholtz's beautiful paper, which is really quite a mathematical gem, I must still say that I think Helmholtz's modifications is rather a retrograde step. It is not so perhaps in the mathematical treatment; but at the same time Helmholtz is perfectly aware of the kind of thing that is meant by viscous consumption of energy. He knows perfectly well that that means, conversion of energy into heat. And in introducing it he is throwing up the sponge as it were, so far as the fight with the dynamical problem is concerned. On the other hand, Sellmeier sticks to it and I think Lommel does.

I got another quarter hundred weight of books on the subject last night. I have not read them all through. I opened one of them this forenoon, and exercised myself over a long mathematical paper. I do not think it will help us very much in the mathematics of the subject. What we want is to try and see if we cannot understand more fully what Sellmeier has done, and what Lommel has done. I see that both stick dimly to the idea that we must account for the loss of energy in the vibration of the particles themselves. That is what I am doing, and we shall never have done with it until we have explained every line in Prof. Rowland's splendid spectrum. If we are tired of it, we can rest, and go at it again.

Lommell and Sellmeier do not go into these multiple vibrations, although they take notice of them.

But they do indicate that we must find some way of distributing the energy without supposing the consumption of it. That is the reason why I do not like Helmholtz's way of introducing the viscous terms. It is very dangerous, in an ideal sense, to introduce them at all. This little bit of viscosity in one part of the system might run away with all our energies long before 50,000 vibrations. If there were any viscosity connected with the moving particle it might be impossible to get a sequence of one-hundred thousand or a million vibrations proceeding from one initial vibration of one vibrator.

What the dynamical problem has to do for us is to show how we can have a system capable of vibrations in itself and acted upon by the luminiferous ether that under ordinary circumstances does not absorb the light in thousands of vibrations. That may be conceived to be the case with transparent bodies; bodies that allow waves to pass through them one hundred feet or a thousand feet, or much greater distances; transparent bodies with exceedingly little absorption. If we take vibrators, then, that will perform their functions in such a way as to give a proper velocity of propagation for light in a highly transparent body and yet which, with a proper modification of the magnitudes of the masses or of the connecting springs will, in certain complex molecules, such as the molecules of some of those compounds that give rise to fluorescence and phosphorescence, take up a large quantity of the energy, so that, perhaps the whole suite of vibrations from a single initiation may be absolutely absorbed and converted into vibrations of a much lower period which will have, lastly, the effect of heating the body, I think we shall see a perfectly clear explanation of absorption without introducing viscous terms at all; and that idea we owe to Sellmeier. I may go for a moment into this subject of arbitrary functions; but perhaps I had better leave it for the present.

I would like, in connection with the idea of explaining absorption and refraction, and lastly, anomalous reflection and dispersion, to just point out as a matter of history, the two names to which this is owing, — Stokes and Sellmeier. I would be glad to be corrected with reference to either, if there is any evidence to the contrary; but so far as I am aware, the very first idea of accounting for absorption by vibrating particles taking up the energy in all modes of natural vibration of their own corresponding to the period of the light trying to pass through, is from Stokes. He taught it to me at a time that I can fix in one way indisputably. I never was at Cambridge once from about June 1852 to May 1865; and it was at Cambridge walking about on the grounds ~~of the~~ of the colleges that I learned it from Stokes. Something was published about it from a letter of mine upon it which was put in a postscript by Kirchhoff to the English translation [Phil. Mag. vol. 20, July 1860] of his own paper on the subject which appeared first in Poggendorff's Annalen, [vol CIX p. 275]. If you have not already read that classical paper of Kirchhoff's, I advise you to look through it at all events, whether you go all through the mathematics or not.

In the postscript you will find the following statement copied from my letter: —

"Prof. Stokes mentioned to me at Cambridge some time ago, probably about ten years, that Prof. Miller had made an experiment testing to a very high degree of accuracy the agreement of the double dark line D of the solar spectrum, with the double bright line constituting the spectrum of the spirit lamp burning salt. I remarked that there must be some physical connection.

between two agencies presenting so marked a characteristic in common. He assented, and said he believed a mechanical explanation of the cause was to be had on some such principle as the following:— Vapour of sodium must possess by its molecular structure a tendency to vibrate in the periods corresponding to the degrees of refrangibility of the double line D. Hence the presence of sodium in a source of light must tend to originate light of that quality. On the other hand, vapour of sodium in an atmosphere round a source must have a great tendency to retain in itself, i.e. to absorb and to have its temperature raised by light from the source, of the precise quality in question. In the atmosphere around the sun, therefore, there must be present vapour of sodium, which, according to the mechanical explanation thus suggested, being particularly opaque for light of that quality, prevents such of it as is emitted from the sun from penetrating to any considerable distance through the surrounding atmosphere. The test of this theory must be had in ascertaining whether or not vapour of sodium has the special absorbing power anticipated. I have the impression that some Frenchman did make this out by experiment, but I can find no reference on the point.

"I am not sure whether Prof Stokes' suggestion of a mechanical theory has ever appeared in print. I have given it in my lectures regularly for many years always pointing out along with it that solar and stellar chemistry were to be studied by investigating terestial substances giving bright lines in the spectra of artificial flames corresponding to the dark lines of the solar and stellar spectra."* [For note see next page.]

What I have read this far is not with reference to the origin of spectrum analysis, of which there is ample historical evidence that it was done before these dates, but the definite point of the dynamics of absorption. There is a hint there of the reaction of the vibrating particles in the luminiferous ether. Sellmeier's first title is to that effect; he takes up exactly that view for explaining absorption. He explains ordinary refraction through the inertia of these particles and he shows how, when the light is nearly of the period corresponding to any of the fundamental periods of the vibrator there will be anomalous dispersion. He gives a mathematical investigation of the subject, not altogether satisfactory, perhaps, but still it seems to me to form a nearer treatment of the thing. Lord Rayleigh, Helmholtz and others have quoted Sellmeier. Lommel begins afresh, I think, but he notices Sellmeier also, so the thing must have originated there, and it seems to me a very important new departure with respect to the dynamical explanation of light.

(NOTE) * [The following is a note appended by Prof. Stokes to his translation of a paper by Kirchhoff in Phil. Mag. Vol. XIX, March 1860, p. 196:– "The remarkable phenomenon discovered by Foucault, and rediscovered and extended by Kirchhoff, that a body may be at the same time a source of light giving out rays of a definite refrangibility, and an absorbing medium extinguishing rays of the same refrangibility which traverse it, seems readily to admit of a dynamical illustration borrowed from sound. We know that a stretched string which on being struck gives out a certain note (suppose its fundamental note) is capable of being thrown into the same state of vibration by aërial

Now let us look at this problem of vibrating particles once more. I have a little question for the ideal arithmetica laboratory. Just try the arithmetical work for this problem for 7 particles. I do not know whether it will work out well or not. I have not the time to do it myself, but perhaps some of you may find the time and be interested enough in the thing, to do it. Take the M's in order, proceeding by ratios of 4 and the C's in order proceeding by differences of 1:

$m_1, m_2, m_3, m_4, m_5, m_6, m_7, = 1, 4, 16, 64, 256, 1024, 4096,$
$C_1, C_2, C_3, C_4, C_5, C_6, C_7, C_8, = 1, 2, 3, 4, 5, 6, 7, 8$

There will be 7 roots to find by trial. I would like to have some of you try to find some of these, if not all also the energy ratios. You will probably find it an advantage in the calculation if you proceed thus: We have $a_1 = \frac{1}{T^2} - 3$, $a_2 = \frac{4}{T^2} - 5$, $a_3 = \frac{16}{T^2} - 7$, $a_4 = \frac{64}{T^2} - 9$, $a_5 = \frac{256}{T^2} - 11$, $a_6 = \frac{1024}{T^2} - 13$, $a_7 = \frac{4096}{T^2} - 15$. You will have to take values of T^2 by trial until you get near a root. The convergence of the continued fraction will be so rapid that you will have very little trouble in getting the largest roots in $\frac{1}{T^2} = Z$. Begin then with the largest root, and proceed downwards; and when several of the a's have become negative, alter the expression so as to keep

---

vibrations corresponding to the same note. Suppose now a portion of space to contain a great number of such stretched strings forming thus the analogue of a "medium". It is evident that such a medium, on being agitated, would give out the note above mentioned; while on the other hand, if that note were sounded in air at a distance, the incident vibrations would throw the strings into vibration and consequently would themselves be gradually extinguished, since otherwise there would be a creation of vis viva. The optical application of this illustration is too obvious to need comment. — G. G. S." H.]

positive quantities. Our standard form is $u_i = a_i - \frac{c_{i+1}^2}{u_{i+1}}$. If a_i is positive, well and good; you will find, at once, u_i a very large number, and so calculate for instance u_1 you may suppose u_4 to be infinite, at the same time supposing u_5 to be infinite, in calculating u_2 u_3. A very few trials will show you how many terms of the continued fraction you must take in order to get u_1 to a certain degree of accuracy. I think, to fix the ideas, and to make the demands for accuracy very moderate, we shall say that our final result shall be within $\frac{1}{10}$th per cent, that is, $\frac{1}{1000}$ of the absolutely true value. That would be correct to 3 decimal places. I do not want to suggest any elaborate arithmetical calculations. Work it out to four places if you like, so as to be quite sure of the third place. Take any value of z you like and calculate; then take another smaller value and you will soon find one that will make $u_1 = 0$. These are the values that we want, the values of z that make $u_1 = 0$. Take smaller values of z, and you will soon find another; take smaller values of z and you will soon find another. By this time you had better begin making the change that I now suggest, viz: take $w_{i-1} = -u_{i-1}$, $b_{i-1} = -a_{i-1}$. As you diminish z, you see that when z becomes less than 3 the a_1 is negative; if $z < \frac{5}{4}$, a_1, a_2 are negative, etc. Instead of these negative values extending up to a_i say substitute positive values $b_1 = -a_1$ etc., at the same time altering the corresponding u's into w's. That will diminish the tendency to negative quantities among the w's; although negative values will sometimes occur. Then proceed backwards from w_1:

The formula will be $w_{i-1} = b_{i-1} - \frac{c_i^2}{w_i}$, or $w_i = \frac{c_i^2}{b_{i-1} - w_{i-1}}$. Put that into the form of a continued fraction if you like but it will be easier to work step by step. Calculate w_i on the supposition that $w_1 = 0$ and u_i on the supposition that $u_{i+1} = \infty$ and equate w_i to $-u_i$ — that is the process. If they are not equal, you must alter z. The value of z that makes them equal must be a root of $U_1 = 0$. In the course of the process you will have the whole formation of the U's or the u's for each root; by multiplying these in order, you have the x's for each particular root, and then you can calculate the energy ratios for each root. We shall then be able to put our formula into numbers; and I feel that I understand it much better when it is in numbers than when it is in a literal form.

I want to show you (jumping ahead a little) the explanation of ordinary refraction. Let us go back to our supposition of spherical shells, if you like — our rude mechanical model. Suppose an enormous number of spherical cavities distributed equally through the space we are concerned with. Let the quantity of ether thus displaced be so exceedingly small in proportion to the whole volume that the elastic action of the residue will not be essentially altered by that. These suppositions are perfectly natural. Now, what is unnatural mechanically, is, let us suppose a massless spherical lining absolutely rigid to this spherical cavity in the luminiferous ether connected by springs — in the first place symmetrical. We shall try afterwards to see if we cannot do something in the way of aeolotropy; but as I have said before I do not see the way out of the difficulties yet. In the meantime, let us suppose this first shell m_1 to be isotropically connected

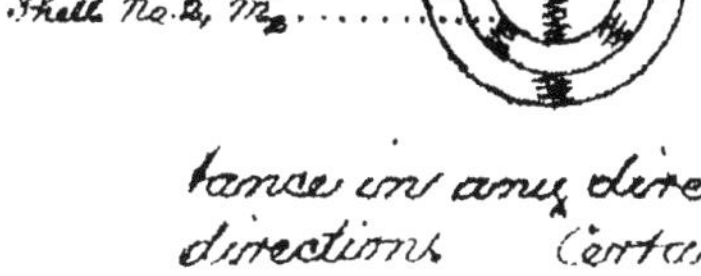

by springs with the rigid shell lining of the spherical cavity in the ether. When I say isotropically connected I mean distinctly this: that if you draw this first shell aside through a certain distance in any direction, the force will be independent of the direction. Certain springs in the drawing — the smallest number would be four — placed around in proper positions will rudely represent the proper connections for us. Similarly, let there be another shell here, m_2, isotropically connected with the outer one; and so on.

This is the simplest mechanical representation we can give of a molecule or an atom, imbedded in the luminiferous ether, unless we suppose the atom to be absolutely hard, which is out of the question. If we pass from this problem to a problem in which we shall have a continuous connection instead of a series of connections of associated particles, we shall be, of course much nearer the reality. But the consideration of a group of particles has great advantage, for we are more familiar with common algebra than with the treatment of partial differential equations of the second order, with coefficients not constant, but functions of the independent variable,—which are the equations we have to deal with if we take a continuous elastic molecule, instead of one made up of masses connected by springs as we have been supposing.

Let us suppose these spherical cavities to be exceedingly small in comparison with the wave length. Practically speaking, we suppose our structure to be infinitely fine grained. That will not in the least degree prevent its doing what we want. The distance also from one such cavity, a series of shells, to another in the luminiferous ether is to be exceedingly small in comparison with the wave length so that the distribution of these molecules through the ether leaves us with a body which is homogeneous when viewed on so coarse a scale as the wave length; but it is, if you like, heterogeneous when viewed with a microscope that will show us the millionth or million millionth of a wave length. This idea has a great advantage over Cauchy's old method in allowing an infinitely fine grainedness of the structure, instead of being forced to suppose that there are only several molecules, ten or twelve to the wave length, as we are obliged to do in getting the explanation of refraction by Cauchy's method.

I wish to show you the effect of molecules of that kind upon the velocity of light passing through the medium. Let $\frac{m_1}{4\pi^2}$ denote the sum of all the masses of shell No. 1 in any volume divided by the volume; let $\frac{m_2}{4\pi^2}$ denote the sum of the masses of No 2 interior shell in any volume divided by the volume; and so on. Or, if you like to say so, let $\frac{m_1}{4\pi^2}$ denote the amount per unit volume of No. 1 shell and so on. We will not put down the equations of motion for all directions, but simply take the equations corresponding to a set of plane waves in which the direction of the vibration is parallel to OX. If we denote by $\frac{\rho}{4\pi^2}$ the density of the vibrating medium, (I am taking $\frac{\rho}{4\pi^2}$ instead of the usual ρ for the reason you know, viz: to get rid of the factor $4\pi^2$ resulting from differentiation) and if n be the rigidity of the luminiferous ether the equation of motion in the ether will be $\frac{\rho}{4\pi^2}\frac{d^2\xi}{dt^2} = n\frac{d^2\xi}{dx^2}$. Let $\frac{\ell}{4\pi^2}$ instead of n denote the rigidity, and the dynamical equation of motion will be $\frac{\rho}{4\pi^2}\frac{d^2\xi}{dt^2} = \frac{\ell}{4\pi^2}\frac{d^2\xi}{dx^2} + C_1(x_1 - \xi)$. I will not go into the formal proof just now, for I am going to take up some dynamics comprehending this when we come to the subject of rotation. We shall suppose that we have gyrostatic fly wheels imbedded in these holes or cavities in the luminiferous ether, and we shall then formally go through the dynamical investigation, and see how it is that we have simply to add to the first equation and expression for the force produced by the springs connecting the lining of the cavity with m_1, which will be $C_1(x_1 - \xi)$.

For waves of period T, we have $\xi = C. \sin 2\pi \left(\frac{x}{\lambda} - \frac{t}{T}\right)$. The second differential coefficient of this with respect to t, x will be $-\frac{4\pi^2}{T^2}\xi$, $-\frac{4\pi^2}{\lambda^2}\xi$ respectively. Therefore our equation becomes $\frac{\rho}{T^2} = \frac{e}{\lambda^2} + C_1\left(1 - \frac{x_1}{\xi}\right)$. Let us find $\frac{T^2}{\lambda^2}$, which is the reciprocal of the velocity of propagation. You may write it $\frac{1}{v^2}$ if you like, or μ^2, the refracting index. We have, $\frac{T^2}{\lambda^2} = \frac{1}{e}\left\{\rho - C_1 T^2\left(1 - \frac{x_1}{\xi}\right)\right\}$. Substitute our value for $-\frac{x_1}{\xi} = \frac{C_1 T^2}{m_1}\left(\frac{\kappa_1^2 R_1}{\kappa_1^2 - T^2} + \frac{\kappa_2^2 R_2}{\kappa_2^2 - T^2} + \dots\right)$ and this becomes

$$\frac{T^2}{\lambda^2} = \frac{1}{e}\left[\rho - C_1 T^2\left\{1 + \frac{C_1 T^2}{m_1}\left(\frac{\kappa_1^2 R_1}{\kappa_1^2 - T^2} + \frac{\kappa_2^2 R_2}{\kappa_2^2 - T^2} + \dots\right)\right\}\right].$$

This is the expression for the square of the refractive index, as it is affected by the presence of molecules arranged in that way. It is too late to go into this for interpretation just now, but I will tell you that if you take T considerably less than κ_1 and very much greater than κ_2, you will get a formula with enough disposable constants to represent the index of refraction by an empirical formula, as it were, which, from what we know, and what Sellmeier and Ketteler have shown, we can accept as ample for representing the refractive index of most transparent substances. We have no means of extending its powers and introducing the effects of these other terms, so that we have a formula which is more than sufficient to give us a mathematical expression of the refrangibility in the case of any transparent body whose refrangibility is reliable.

We shall look into this a little more, and I will point out some of the applications to anomalous dispersion. We must think a good deal of what can become of vibrations in a system of that kind when the period of the vibration of the luminiferous ether is approximately equal to any one of the fundamental periods that the system could have were the shell lining in the ether held absolutely at rest.

Lecture X.

We shall now think a little about the propagation of waves with a view to the question, what is the result as regards waves at a distance from the source, those at the source being discontinuous. In the first place, we will take our expression for a plane wave. The expression in our formulas showing diminution of amplitude at a distance from a source does not have an effect when we come to consider plane waves. So we just take the simple expression for plane harmonic waves propagated along the axis of y with velocity v; $\xi = a \cos \frac{2\pi}{\lambda}(y - vt)$.

Let us consider this question, what is the work done per period by the elastic force in any plane perpendicular to the line of propagation of the wave. We shall think of the answer to that question with the view to the consideration of the possibility of a series of waves penetrating through space previously quiescent. Suppose I draw a straight line here for the lines of propagation and

Let this curve represent a succession of waves travelling from left to right and penetrating into space previously quiescent. Take a plane perpendicular to the line of propagation of the waves, and think of the work done by the elastic solid upon one side of this plane upon the elastic solid on the other side, in the course of a period of the vibration. We shall take an expression for the tangential force T of the elastic solid. I am not adhering to our old notation of S, T, U, P, Q, R. We shall virtually investigate here the formula for the propagation of the wave independently of our general formula in three dimensions. Take T to denote the tangential force of the elastic medium on the one side of this plane; the direction of the arrow head which I draw being that direction in which the medium on the left pulls the medium on the right. I put infinitely near that in the medium on the left another arrow head. I cannot do that actually; it is an easy thing to understand, but not a practical thing to do. Imagine for the moment a split in the medium caused by this plane; and imagine the medium on the left taken away, and that you act upon this plane with the same force as in the continuous propagation of waves. The medium upon the left acts in this way upon the plane – that is an easy enough conception. I correctly represent that by an arrow head pointing up infinitely near to the plane on the right hand side, and an arrow head on the left pointing down. The displacement of the medium is determined by a distortion from a square figure to an oblique figure and there is no inconsistency in putting into this little diagram an exaggeration of the obliquity so as to show the direction of it: 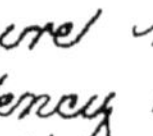. The force required to do that is clearly upward on the right and downward on the left

Let us consider now the work done by that force. Calling ξ the displacement of a particle from its mean position, $T.\dot{\xi}$ is the work done by that tangential force per unit of time. $\frac{d\xi}{dy}$ is the shearing strain experienced in the medium, so that $n\frac{d\xi}{dy}=-T$. In this particular position which we have taken, ξ increases with y, so that the minus sign is correct according to the arrow heads.

Let there be simple harmonic waves propagated from left to right with velocity v. This is the expression for it [indicating $\xi = a\cos\frac{2\pi}{\lambda}(y-vt)$]. Hence, $\dot{\xi}=\frac{2\pi}{\lambda}va\sin q$, $\frac{d\xi}{dy}=-\frac{2\pi}{\lambda}a\sin q$; and the rate of doing work is $\frac{4\pi^2}{\lambda^2}a^2vn\sin^2 q$. That is the rate at which this plane, working on the elastic solid on the right hand side of it does work ("per unit area of the plane" understood). Multiply this by dt and integrate through a period $T=\frac{\lambda}{v}$. Now $\int_0^T \sin^2 q\,dt=\int_0^T \frac{1}{2}(1-\cos 2q)\,dt=\int_0^T \frac{1}{2}\,dt=\frac{1}{2}T$. The rate of doing work, then, per period, is $\frac{2\pi^2}{\lambda^2}a^2vnT=\frac{2\pi^2a^2n}{\lambda}$.

If it is possible for a set of waves to advance into space previously undisturbed, then it is certain that the work done per period must be equal to the energy in the medium per wave length. Let us then work out the energy per wave length.

It is easily proved that the energy is half potential of elastic stress, and half kinetic energy; and it will shorten the matter, simply to calculate the kinetic energy and double it, taking that as the energy in the medium per wave length. In our notation of yesterday, we took $\frac{\rho}{4\pi^2}$ as the density. Multiply this by dy, to get the mass of an infinitessimal portion (per unit of area in the plane of the wave). The kinetic energy of this mass is $\frac{1}{2}\frac{\rho}{4\pi^2}dy\,\dot{\xi}^2=\frac{1}{2}\frac{\rho v^2a^2}{\lambda^2}\sin^2 q.\,dy$. Integrating this through a wave length ($\int_0^\lambda \cos 2q\,dy=0$), and doubling it so as to get the whole energy we have $\frac{1}{2}\frac{\rho v^2 a^2}{\lambda}$. Compare that with the work done per period, viz: $\frac{1}{2}\frac{a^2}{\lambda}l$. if $\frac{l}{4\pi^2}$ be as yesterday the rigidity instead of $\underline{n}$. That gives us correctly the velocity, $v=\sqrt{\frac{l}{\rho}}$. Thus the work done per period is equal to the energy per wave length.

We must not infer from this that it is possible for a discontinuous series of waves to be propagated into the elastic medium, pr eviously quiescent. But if this did not verify, it would be impossible to have such a series of waves propagated forward without change of form into a medium previously quiescent. I wanted to verify that case, because for a moment, we shall alter to a case in which this is not verified; that is to say, when we put in our molecules. In that case the work done per period is less than the energy in the medium per wave length, and therefore it is impossible for the waves to advance without change of form.

Before we go on to that, let us stay a little longer in an undisturbed elastic solid, and look at the well known solution by discontinuous functions. The equation of motion is $\rho\frac{d^2\xi}{dt^2} = l\frac{d^2\xi}{dy^2}$. Although I said I would not formally prove this now, it is in reality proved by our old equation $\rho\frac{d^2\xi}{dt^2} = l\nabla^2\xi$. — I took the liberty of asking Prof. Ball two days ago whether he had a name for this symbol ∇^2; and he has mentioned to me nabla, a humorous suggestion of Maxwell's. It is the name of an Egyptian harp which was of that shape. I do not know that it is a bad name for it. Laplacian I do not like for several reasons both historical and phonetical.

I should have told you that this is the case of a plane wave propagated in the direction of OY, with the plane of the wave parallel to XZ; for which case, nabla of ξ becomes simply $\frac{d^2\xi}{dy^2}$. The time honored solution of this equation is $\xi = f(y-vt)+F(y+vt)$, where f and F are arbitrary functions. You can verify that by differentiation. This solution in arbitrary functions proves that a discontinuous series is possible; and knowing that a discontinuous series is possible, you could tell without working it out, that the work done per period by the medium on the one side of the plane which you take perpendicular to the line of propagation must be equal to the energy of the medium per wave length.

Before passing on to the energy solution for the case in which we have attached molecules in which this equality of energy and work does not hold with the result that you cannot get the discontinuous series. I want to suggest another elementary exercise for the anticipated arithmetical laboratory. It is to illustrate the propagation of waves in a medium in which the velocity is not independent of the wave length; and to contrast that with the propagation of waves when the velocity is independent of the wave length in order that you may feel for yourselves what these two or three symbols show us, but which we need to look at from a good many points of view before we can make it our own and understand it thoroughly. To realize that this equation $E = F$ gives us constant velocities for all wave lengths and that constant velocities for all wave lengths implies this equation and to see that that goes along with the propagation of a discontinuous pulsation without change of figure, or a discontinuous succession of pulsations without change of character; I want an illustration of it, and also of the case in which the conditions of constancy of velocity for different wave lengths are not fulfilled.

I ask you first to notice the formula $S = \frac{\frac{1}{2}(1-e^2)}{1-2e\cos q+e^2} = \frac{1}{2} + e\cos q + e^2\cos 2q + \ldots$ which is familiar to all mathematical readers as leading up to Fourier's harmonic series of sines and cosines. Poisson and others make this series the foundation of a demonstration of Fourier's theorem. It is proved by taking $2\cos q = e^{iq}+e^{-iq}$ and resolving into partial fractions. If $e<1$ the series is convergent; when $e=1$ it ceases to converge. If we take $q = \frac{2\pi}{\lambda}(y-vt)$ and draw the curve whose dependent coordinate is S, what have we? Take $t=0$ and measure off lengths from the origin $y=\lambda$, $2\lambda, \ldots$. The curve represented will be this (heavy curve)

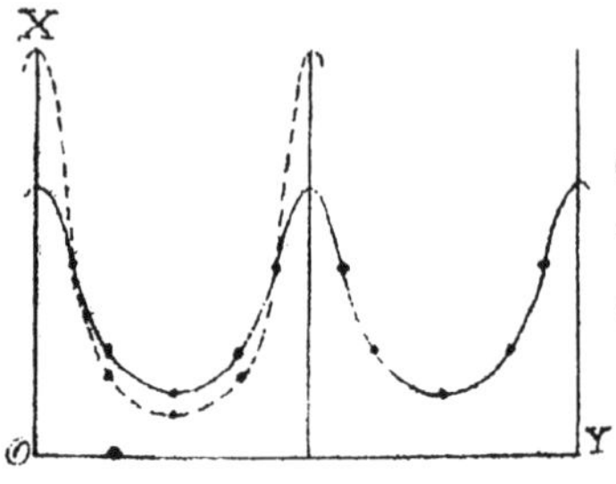

The heavy curve is, $2S = \frac{4}{5-3\cos 2\pi y}$ $(\lambda=1, e=\frac{1}{3})$.
It is here drawn by the points $(y,S)=(0,1),(\frac{1}{8},\frac{7}{10}),(\frac{1}{4},\frac{4}{10}),(\frac{1}{2},\frac{1}{4})$ etc.
The dotted curve is $2S = \frac{3}{5-4\cos 2\pi y}$ $(\lambda=1, e=\frac{1}{2})$
It is here drawn by the points $(y,S)=(0,\frac{3}{2}),(\frac{1}{8},\frac{7}{10}),(\frac{1}{4},\frac{3}{10}); (\frac{1}{2},\frac{1}{6})$ etc.

If you take any other value of t than zero, you merely shift the curve as it were along the axis of y. I want the arithmetical laboratory to work this out and give a graphic representation of the periodic curve for one or two different values of e if you like. Perhaps a dozen equal difference values of y will be more than enough to trace a good curve corresponding to this equation. The particular numerical case that I am going to suggest is one in which the curve will be more like this second curve which I draw (dotted curve); it is much steeper and comes down more nearly to zero. Take the extreme case of $e=1$, and what happens? S is infinitely great for y infinitely small, and infinitely small for all other values of $y<\lambda$. I suggest to work this out for $e=(.1)^{1/10}$. The coefficient in the tenth term will be a tenth part of the coefficient in the first term. On the other hand, if you take e something smaller, say $\frac{1}{2}$, the series will converge so rapidly that long befor the tenth term occurs the terms will be too small for any calculation that I would recommend to the arithmetical laboratory.

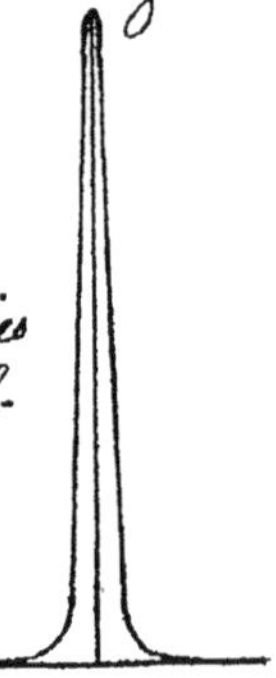

There will be no necessity to calculate the

terms of this series if you have no other object than to trace this curve. Take the curve $2S = \frac{1-e^2}{1-2e\cos\varphi+e^2}$. For $y=0$, $\varphi=0$ and $2S = \frac{1+e}{1-e}$; for $y=\frac{1}{2}\lambda$, $\varphi=\pi$, $2S = \frac{1-e}{1+e}$. Now, the tenth root of .1 is nearly .8, and corresponding to this value of e the maximum value of $2S$ is 9, and the minimum value is $\frac{1}{9}$; so that the case I have suggested makes the height at the origin about 81 times the minimum height here. If you want to get a still more telling expression, take a value of e still nearer unity. This problem is worth working out in itself, and I would advise those also who have time to read Poisson's and Cauchy's great papers in connection with it [Poisson: Mémoire sur la theorie des ondes. Paris, Mém. Acad. Sci. I, 1816, pp 71-186; Annal de Chémie, V., 1817, pp 122-142. Cauchy, Mémoire sur la théorie de la propagation des ondes à la surface d'un fluide pesant d'une profondeur indéfinie [1815]. Paris, Mém. Sav. É trang. I, 1827, pp 3-312] Those papers are exceedingly fine pieces of paper mathematics, but they are very strong. You might have the hydrodynamical beginnings presented much more fascinatingly. If you know the theory of deep sea waves, well and good, then take Poisson and Cauchy. Those who do not know the theory of deep sea waves may read it up in elementary books. The best book I know is Lamb's Hydrodynamics. The great struggle of 1815 (that is not the same idea as La grande guerre de 1815) was who was to rule the waves, Cauchy or Poisson. Their two memoirs seem to me of very nearly equal merit. I have no doubt the judges had some particular reason for giving the award to Cauchy, but Poisson's paper is splendid. I can see that the two writers respected each other very much and I suppose each thought the other's work as good as his own.

I want to know a little more myself whether or not we can get from this series a graphic representation of the effect of a single disturbance at sea —

such a disturbance as that of throwing a stone in a deep sea. I believe there are quite valid solutions to be obtained, but there are difficulties, such as questions of convergency, and so on. That is the problem I believe they did; it constitutes the largest part of their papers; but they go into it in the high analytical style of letting the initial condition be quite arbitrarily chosen. Every portion of an infinite area of water is started initially with a stated infinitessimal displacements from the level and a stated velocity up and down from the level and the inquiry is, what will be the result? The solution of this constitutes the problem; but it is obvious that you h ve the solution of that problem from the more elementary problem, what is the result of an infinitessimal displacement at a single point, which may just as well be produced by throwing in a stone as in any other way. Let a solid, say, cause a depression in any place, the velocity of the solid performing the part of giving velocity to the particles of water and then suddenly consider the solid annulled. The same thing in two dimensions is exceedingly simple. Take for example, waves in an infinitely deep canal with vertical sides. Take a sudden disturbance in the canal equal all along the breadth of the canal and inquire what will the result be. That leads toward an understanding of Cauchy's and Poisson's solution and I think it would repay any one who is inclined to go into the subject to work it out theoretically and make graphic representations. Poisson and Cauchy only give figures and do not give graphical representations.

I am going to suggest to the arithmetical laboratory to take the case of $\underline{v}$ the velocity dependent on the wave length. Let us take this as the arithmetical problem: The curve to be drawn for $S = \frac{1}{2} + c \cos q_1 + c^2 \cos 2q_2 + c^3 \cos 3q_3 + \ldots$ where $q_i = \frac{2\pi}{\lambda}(y - v_i t)$. For a particular case take $v_i = \frac{1}{\sqrt{i}}$, and calculate the curve corresponding to any values you please of t. First give t a small value corresponding say to the time when you have a velocity 1. You might for example take for the first case $t = \frac{1}{4}$. You will find the result will be a shifting of this curve to one side about a quarter of a wave length. Then try the cases $t = 1, 2$, etc., calculating enough of the terms of this series to give you a fairly representative curve. It is not a thing that can be done quickly. It is worth putting a good deal of labor upon, and I mean myself to do it, putting the calculation into the hands of some of my assistants who will be glad to work out what I think will be a somewhat valuable representation of this interesting property.

We are going to take our molecules again and put them in the ether and look at the question a little more, what is the velocity of propagation under some suppositions which we shall make as to the masses of these attached molecules, and how much it will modify the velocity of propagation from what it would be if there were no molecules. Then we shall look at the matter, with no more work to do upon it with respect to the question of the work done upon a plane perpendicular to the line of propagation, and we shall see that the energy per wave length is much greater than the work done per period and that therefore it is impossible under these conditions for waves to spread into space previously occupied by quiescent matter.

You will find in Lord Rayleigh's book on sound the question of the work done per period and the energy per wave length gone into and the application of this principle with respect to the possibility of independent suites of waves travelling without change of form is thoroughly pointed out.

To-morrow we shall consider a piece of work that looks to the difference of velocity of propagation in different directions in an aeolotropic elastic solid for the foundation of the explanation of double refraction on the pure elastic solid idea. The thing is quite familiar to many of you, no doubt, and you also know that it is a failure in regard to the explanation of the propagation of light in biaxial crystals. It is, however, an important piece of physical dynamics, and I shall touch upon it a little, and try to show it in as simple a point of view as I can.

Now for our proper molecular question. The distance from cavity to cavity in the ether is to be exceedingly small in comparison with the wave length, and the diameter of each cavity is to be exceedingly small in comparison with the distance from cavity to cavity. Let the lining of the cavity be an ideally absolutely rigid massless shell. Let the next shell be an absolutely rigid shell of mass $\frac{m_1}{4\pi^2}$. I represent the thing as if we had just two of these shells and a solid nucleus.

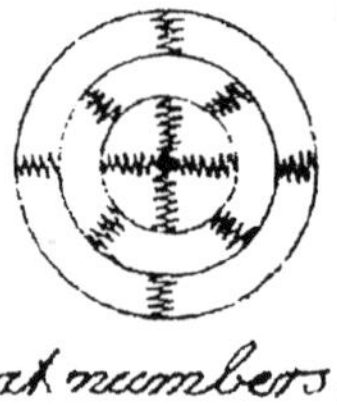

The enormous mass of the matter of the grosser kind which exists in the luminiferous ether or even of such a comparatively non-dense body as air, would bring us at once to very great numbers in respect to the masses which we will suppose inside this cavity in comparison, with the masses of comparable bulks of the luminiferous ether. If there is time to-morrow, we shall look a little to the possible suppositions as to the density of the luminiferous ether.

what limits of greatness or smallness are conceivable in respect to it. At present we have enough to go upon to show that even in air of ordinary density, the mass of air per cubic centimeter must be enormously great in comparison with the mass of the luminiferous ether per cubic centimeter. We must have something enormously massive in the interior of these cavities. We shall think a good deal of this yet to try and find how it is we can have the large quantity of energy that is necessary to account for the heating of a body such as water by the passage of light through it; or for the phosphorescence of a body which is luminous for several days after it has been excited by light. I do not think we shall have the slightest difficulty in explaining these things. These are not the difficulties. The difficulties of the wave theory of light are difficulties which do not strike the popular imagination at all. These are the difficulties of accounting for polarization by reflection with the right amount of light reflected and for double refraction. With the phenomena we have no difficulty whatever; the great difficulty in respect to the wave theory of light is to bring out the proper quantities in these effects.

People seem to think the luminiferous ether a fanciful idea. I wish to give another illustration besides shoemaker's wax; I ask you to think of glycerine. Glycerine is a substance without any coarse structure; it is molecularly fine. Glycerine takes its level if you pour it out, as accurately as water or mercury does, yet if you suddenly change its shape, it springs back. Many of you may remember Maxwell's beautiful experiments in which the effects of strain on polarized light was shown in a liquid or rather body which, if you give it time takes its level absolutely, and yet if you strike it quickly, it springs back, Canada Balsam was the substance. Any of you who have any desire to do so may try the experiment. Put a stick in Canada Balsam, get the proper polarizing appliances, make a sudden turn with the stick and you will see the optical effects of double refraction produced and gradually fading away.

This is a digression from my subject, but I do not want to part from you without letting you know all I can in the way of helping you think of the luminiferous ether as a reality, and that we are speaking of real bodies and that this is not a mystification of the mind.

There is no difficulty in explaining the energy required for heating a body by radiant heat passing through it, nor how it is that it sometimes comes out as visible light and, it may be not so fast but that we may get light for two or three days. All these properties, remarkable as they are, seem to come out as a matter of course from the dynamical consideration. So much so that any one not knowing these phenomena would have discovered them on working out these things dynamically. He would discovered anomalous dispersion, fluorescence, phosphorescence, and the phosphorescence corresponding to lower periods, consisting in the heating of a body and then afterwards giving that out as heat. All these phenomena might have been discovered by dynamics; and the dynamical treatment that discovers what is afterwards verified by experiment is a very competent piece of dynamics.

I speak with confidence in this subject because it is a mater of fact. I am ashamed to say that I never heard of anomalous dispersion until after I found it lurking in the formulas. I said to myself, "These formulas would imply that, and I never have heard of it." And when I looked into the matter I found to my shame that a thing which had been known by others for eight or ten years I had not known until I found it in the dynamics.

Take our formula which we had yesterday, $\frac{\rho}{4\pi^2}\frac{d^2\xi}{dt^2} = \frac{e}{4\pi^2}\frac{d^2\xi}{dy^2} - C_1(\xi - x_1)$ and try this with some simple harmonic motion, $\xi = a \sin \frac{2\pi}{\lambda}\left(\frac{y}{\lambda} - \frac{t}{T}\right)$. From this we find $\frac{\rho}{T^2} = \frac{e}{\lambda^2} + C_1\left(1 - \frac{x_1}{\xi}\right)$, which solved for the refractive index gives $\frac{eT^2}{\rho\lambda^2} = 1 - \frac{C_1}{\rho}\left(1 - \frac{x_1}{\xi}\right)T^2$. I want to take our formula for $-\frac{x_1}{\xi}$ in order to find out what position the period T may have among the fundamental periods of the vibrator on the supposition of the bounding shell held fixed, to give us a good reasonable explanation of dispersion, something in accord with the facts of observation with respect to the difference of velocity for different periods. I will not introduce the energy ratios just now, because we have not time to use them, and I will just take $-\frac{x_1}{\xi} = \left(\frac{q_1}{\kappa_1^2 - T^2} + \frac{q_2}{\kappa_2^2 - T^2} + \ldots\right)T^2$.

In a medium which is denser than the luminiferous ether, the refractive index is always greater, the velocity smaller. If T were less than the smallest of the fundamental periods $-\frac{x_1}{\xi}$ would be positive and the refractive index would be less than unity. But in all known cases the refractive index is greater than unity; therefore $-\frac{x_1}{\xi}$ must be negative. Take then this formula: $-\frac{x_1}{\xi} = \left(\frac{-q_1}{T^2 - \kappa_1^2} + \frac{q_2}{\kappa_2^2 - T^2} + \ldots\right)T^2$. In other words, we shall suppose the period T to be intermediate between the smallest and the next to the smallest of the fundamental periods, κ_1, κ_2. I want to see if we can get out of this a formula which will cover a range, including all light from the highest ultra-violet photographic light of about ½ the wave length of sodium light down to the lowest we know of which is the radiant heat from a Leslie cube with a wave length that I hear from Prof. Langley since I spoke on the subject about a week ago of about $\frac{1}{1000}$ of a centimeter or 17 times the wave length of sodium light. That will be a range of about 40 : 1. The highest chemical light has a period about $\frac{1}{40}$ part of the period of the lowest invisible radiation of a radiant heat that has yet been experimented upon.

It is conceivingly possible that there are some mediums throughout every part of that range for which there are no anomalous dispersions. I think it is almost certain that for rock salt in the lower part of the range, there are no anomalous dispersions at all. In fact Langley's experiments in radiant heat are made with rock salt; and in all experiments made with rock salt, it seems as if little or no radiant heat is absorbed by it. At all events we could not be satisfied unless we can show that this kind of supposition will account for dispersion through a range of period from one to forty. It is obvious that if we are to have continuous refraction without anomalous dispersion through a wide range, T must not exceed another period. κ_2 must then be 40 times as great as κ_1.

If we substitute our value of $-\frac{x_1}{\xi c_1}$ and work it out algebraically, we shall find $\mu^2 = 1 + \frac{c_1}{\rho}\left\{q_1\kappa_1^2 - (1-q_1)T^2 + q_1\kappa_1^2\left(\frac{\kappa_1^2}{T^2} + \frac{\kappa_1^4}{T^4} + ..\right) - (\text{terms involving } q_2, q_3 \ldots)\right\}$. q_1 is essentially less than unity. To agree with anything we know, $q_1\kappa_1^2$ must be large in comparison with $(1-q_1)T^2$. This term $(1-q_1)$ must be so small that an exceedingly large multiplication of it (for instance corresponding say to the range from the sodium D line to the lowest radiant heat $= 17^2$), must not have any very serious effect; it may be a correction upon the other terms but it must be small. We have here two disposable constants q_1, κ_1. I shall look at this a little more carefully tomorrow, and think perhaps, of numerical solutions of our continued fraction and how it is we can suppose q_1 very nearly unity – I think within $\frac{1}{100,000}$ of unity.

What will that mean? That the springs between the rigid shell lining and m_1 are so strong that the static displacement of the lining (with the center of mass held at rest) makes the displacement of m_1 very nearly equal to the displacement of the lining. If you the lining to one side, m_1 will be displaced somewhat less than the lining; m_2 somewhat less than m_1; and so on. If we suppose the displacement of the lining to be exceedingly little greater than the displacement of m_1 we get an expression that will be applicable to this case.

We shall study this a little more to-morrow and think of what we can make of the graver and graver modes. Although I cannot promise you much light upon it, we must think of it in connection with this question: Suppose you give a slight shock to the lining and hold it fixed. Then sometime after give another slight shock to the lining and hold it fixed, and so on; what will be the disposition of the energy? How will it creep inwards among the masses? I think that our arithmetical work will help us to see our way to the answer to some of these questions; and through them we shall be able to form perfectly definite dynamical notions of fluorescence and phosphorescence and anomalous dispersion.

Lecture XI.

We shall now take up the subject of an elastic solid which is not isotropic. As I said yesterday, we do not find the consideration of the homogeneous elastic solid satisfactory or successful for explaining the properties of crystals with reference to light. It is, however, to my mind quite essential that we should understand all that is to be known about homogeneous elastic solids and waves in them, in order that we may contrast waves of light in a crystal with waves in a homogeneous elastic solid. It is one of the interesting theories in physical science to know the possibilities of aeolotropy.

Aeolotropy is in analogy with Cauchy's word isotropy which means equal properties in all directions. The formation of a word to represent that which is not isotropic was a question of some interest to those who had to speak of these subjects. I see the Germans have adopted the term anisotropy. Thus they would have us say: "An anisotropic solid is not an isotropic solid;" and this jangle between the prefix <u>an</u> and the article <u>an</u>, if nothing else, would prevent us from adopting that method of distinguishing a non-isotropic solid from one which is. I consulted Prof. Cushing and we had a good deal of talk over the subject. He gave me several charming Greek illustrations and we wound up on the word aeolotropy. Prof. Cushing pointed out that <u>aeolos</u> means variegated, and it is interesting that the Greeks used the word variegated in respect to shape, color, and time. There is no doubt of the classical propriety of the word and it has turned out very convenient in science. That which is different in different ~~in different~~ directions, or is variegated according to direction is called aeolotropy.

The consequences of aeolotropy upon the motion of waves or the equilibrium of particles in an elastic solid is an exceedingly interesting and a fundamental subject in physical science: so that there is no apology in making it a subject here except, perhaps, that it is too well known. On that account I shall be very brief and merely call attention to two or three fundamental points. I am going to take up presently, as a branch of molar dynamics, the actual propagation of a wave; and on the mathematical investigation, I am going to give you nothing but what is true of the propagation of a plane wave in an elastic solid, not limited to any particular condition of aeolotropy, but in an elastic solid which has aeolotropy of the most general kind.

Before doing that which is strictly a problem of continuous or molar dynamics, I want to touch upon the somewhat cloud-land molecular beginning of the subject and refer you back to the old papers of Navier and Poisson, in which the laws of equilibrium or motion of an elastic solid were worked out from the consideration of points mutually influencing one another with forces given functions of the distance. There can be no doubt of the mathematical validity of investigations of that kind and of their interest in connection with molecular views of matter; but we have long passed away from the stage in which Father Boscovich is accepted as being the originator of a correct representation of the ultimate nature of matter and force. Still, there is a never ending interest in the definite mathematical problem of the equilibrium of motion of a set of points endowed with inertia and mutually acting upon one another with any given force. We cannot but be conscious of the one grand application of that problem to what used to be called physical astronomy but which is more properly called dynamical astronomy, or the motions of the heavenly bodies. We have cases in which we have these motions instead of the approximate equilibriums or infinitessimal motions which form the subject of the special molecular dynamics that I am now alluding to.

All writers who have worked upon this subject have come upon a certain definite relation or set of relations between moduluses of elasticity which seemed to them essential to the hypothesis that matter consists of particles acting upon one another with mutual forces, and that the elasticity of a solid is the manifestation of the force required to hold the particles displaced infinitessimally from the position in which the mutual forces will balance. This, which is sometimes called Navier's relation sometimes Poisson's relation, and in connection with which we have the well known Poisson's ratio, I want to show you is not an essential of the hypothesis in question. The result for the case of an isotropic body is a most interesting one: doubtless most of you know it; it is in Thomson & Tait, and I suppose in every elementary book upon the subject. I will just repeat it.

An isotropic solid, according to Navier's or Poissons theory, would fulfil the following condition: if a column of it were pulled lengthwise, the lateral dimensions would be shortened by one half the proportion that the length is added to: and the area of a cross section would therefore be reduced in the double ratio or would be a quarter of the elongation. Stokes called attention to the viciousness of this conclusion as a practical matter in respect to the realities of elastic solids. He pointed out that jelly and india-rubber and the like, instead of exhibiting lateral shrinkage to the extent of one-quarter of the elongation gave only enough shrinkage to cause no reduction in value at all. That is to say, india-rubber and such bodies vary the area of the cross section in inverse proportion to the elongation so that the product of the length into the area of the cross section may remain constant.

Stokes also referred to a promise that I made I think it was in the year 1856, to the effect that out of matter fulfilling Poisson's condition a model may be made of an elastic solid, which when the scale of parts is sufficiently reduced will be a homogeneous elastic solid not fulfilling Poissons condition. Stokes refers to that promise of mine which was made, very nearly 30 years ago. I propose this moment to fulfill it never having done so before. It is a very simple affair.

Let this box represent a rectangular parallelopipedon. The kind of elastic model I am going to suppose is this: a set of particles arranged symmetrically in rectangular order and connected by springs in a certain definite way. I am going to show you that we can connect 8 particles in the interior of an elastic solid with a sufficient number of springs to fulfil the condition of giving 18 independent moduluses; then by transforming the coordinates from a portion of the solid made up in this partially symmetrical manner with respect to the axes to a portion of the solid taken at random, we get the celebrated 21 coefficients, or moduluses of Green's. I suppose you all know that Green took a short cut to the truth; he did not go into the physics of the thing at all, but simply took the general quadratic expression for energy with its 21 independent coefficients as the most general supposition that can be made with regard to an elastic solid.

To make a model of a solid having the 21 independent coefficients of Green's theory, think of how many disposable springs we have with which to connect

8 different particles. Let them be connected first along the 12 edges of the parallelopipedon. That clearly will not be sufficient to give any rigidity of figure whatever, so far as distortions in the principle planes are concerned. These 12 springs connecting in this way these 8 particles would give a resistance to elongation in the directions of the axes; but no restance whatever to obliquity; you could easily change it from rectangular into an oblique figure. What then must we have to give resistance to obliquity? We can connect coplanar particles diagonally. We have in the first place, the two diagonals in each face although the two will virtually count as but one; and then we have the four body diagonals.

Now let me see how many disposables we have got. Remark that each edge is common to four parallelopipedons. I am not going to duplicate our points. We might do it, I suppose, and build up our elastic solid in that way; but I would rather suppose these to be 8 particles of which I show the connections among themselves but not the similar connections with their neighbors in other directions. Each edge being common to four parallelopipedons we have only a quarter of the number of edge springs independently available.* Therefore we have virtually three disposables from the edge springs. Each force is common to two parallelopipedons; therefore from the two diagonals in each face we have only one disposable, making in the six faces six disposables. We have the four body diagonals not common to any other parallelopipedons and therefore four disposables from them. We have now 13 disposables

* [In other words, the eight connected particles forming a model of the whole medium; the bodily translation of the spring connections in the medium must give the same model merely translated parallel springs must be equal. H.]

and I want two more. These are the two proportions of the figure, the ratios of the three principle edges. These 15 disposables are all we can absolutely get by springs arranged in this manner. We want three more, observe, in order to make up the eighteen. Now I thought of the way to get the three is this.

Navier's and Poisson's theory gave an essential relation between the compressibility and the rigidity and made an incompressible elastic solid impossible. It is curious that they did not notice that jelly is practically incompressible. It is a wonder that they did not try it, and see that it did not fulfil Poisson's ratio. Their mistake was due to the vicious habit in those days of not using examples and diagrams. In the Mécanique Céleste you find no diagrams, nor in Lagrange, nor in Poisson's splendid memoir on Waves. I think I refer to it in Thomson and Tait, that if Lagrange had been in the habit of making diagrams, he never would have given out the proposition that whereas a ball is in stable equilibrium in the bottom of an elliptical dish cover, turned with the mouth up, it is in unstable equilibrium in the bottom of a cylindrical bowl. If they had been in the habit of using diagrams and thinking of their symbols more than they were, Lagrange would never have fallen into that mistake; and Poisson and Navier would have found that jelly is enormously more non-compressible than their theory would make it.

What I want is to get a condition of compressibility. I must find some other disposables that will enable me to give any compressibility I please in the case of an isotropic solid. Take our 15 disposables, and reduce them down to the case of an isotropic solid and we find that an isotropic solid made up in this way will have an absolutely definite compressibility; we can not make the compressibility what we please. We must put in something that can make it incompressible or have any compressibility we please, so that we can make out the fit for either cork or india-rubber, the extremes of natural bodies. I must confess that it is the most difficult thing on it, after I got the idea, to run a cord twice around the 12 edges of a parallelopipedon. Here you see the problem solved by these cords running around the edges of this parallelopipedon through a ring in each of the 8 corners. It cannot be done symmetrically, that is a mathematical proposition – at least I suppose it is. But just follow the cord and we will find how to do it. In fact I am finding out how to do it again in a certain way myself. The following is an arrangement of the corners along the cord in succession given by their coordinates:

(000) (001) (011) (010) (000) (001) (011) (010) (000) (100) (110) (010) (110) (111) (011) (111) (101) (001) (101) (111) (110) (100) (101) (100).

There are plenty of other ways of doing it, but this is one way. We have got a cord thrice through each of these 8 points and the thing is done.

Suppose, for example, we wanted to make a condition of incompressibility; let this be an inextensible cord and thing is done. But some one may say that we have not done it without introducing a flexible body. I will not admit any objection to this being a purely mechanical model because we have that inextensible and perfectly flexible cord running around through hooks; but it is interesting to notice that we can do it without introducing a flexible body at all. We can do it with nothing but rigid bodies. Instead of a cord passing through rings take wire, with bell cranks everywhere where that cord bends around a corner and the thing is done. Thus by proper bell cranks fixed at the corners, and inextensible cords connecting them you have fulfilled the condition that the sum of the 12 edges shall be constant, which in the circumstance of being infinitely nearly a rectanguïar figure in all the distortions that we have; is equivalent to saying that the volume is constant.

To see that this gives us the requisite disposables, let the portions of the cords along the 12 edges be of different elasticities. That gives us 3 disposables, each edge being common to 4 others. To speak in mechanical language, let us connect the bell cranks by springs of different strengths in the directions of the three principal edges. When the body is in equilibrium, there is no pull on the springs. Each one of the 16 different independent springs that we have now got will be called into play by a perfectly general displacement of infinitely small amount. We have 18 available quantities which will make by solution of linear equations the required 18 moduluses. Then, as I have said, with the transformation of our solid to rectangular axes in any direction, you have a solid fulfilling Greens conditions in the most general way.

Now, observe, Poisson and Navier give us the means of making a bell crank, although they do not give us means of making a jelly. They gives us the means of making an elastic zig-zag spring. We can take solids fulfilling their theory and make bell cranks and springs out of them. Put these together: make the parts small enough and the number of them great enough, and you have a homogeneous elastic solid constructed out of parts satisfying Poisson's law, which, as a whole does not satisfy it.

Although the molecular constitution of solids supposed in these remarks and mechanically illustrated in our model is not to be accepted as true in nature still the construction of a mechanical model of this kind is undoubtedly very instructive, and we could not be satisfied unless we could see our way to make a model with the 18 independent moduluses. My object is to show how to make a mechanical model which shall fulfil the conditions required in the physical phenomena that we are considering, whatever they may be. At the time when we are considering the phenomenon of elasticity in solids, I want to show a model of that. At another time, when we have vibrations of light to consider, I want to show a model of the action exhibited in that phenomenon. We want to understand the whole about it: we only understand a part. It seems to me that the test of "Do we or not understand a particular subject in physics?" is, "Can we make a mechanical model of it?" I have an immense admiration for Maxwell's mechanical model of electro-magnetic induction. He makes a model that does all the wonderful things that electricity does in inducing currents, etc., and there can be no doubt that a mechanical model of that kind is immensely instructive and is a step towards a definite mechanical theory of electro-magnetism.

I want now to go through a piece of mathematical work, which, so far as I know, is not given anywhere except in the articles on Elasticity in the Encyclopedia Britannica, although nearly the same was given first by Green. Green investigates the propagation of a wave in an elastic solid, but not in a perfectly general elastic solid. He gave it a certain degree of symmetry before he began this investigation; but he need not have done so. The investigation would have been almost letter for letter the same if he had made it before instead of after introducing the effects of symmetry. The investigation I refer to is that of the propagation of a plane wave,— the most general possible kind of a plane wave. Green does it the same way that I am doing it, but with this difference that I make absolutely no supposition regarding simplification by symmetrical qualities of the solid.

A plane wave in a homogeneous elastic solid is a motion in which every line of particles in a plane parallel to one fixed plane experiences simply a motion of translation — but a motion differing from the motions of particles in planes parallel to the same. Let OX be perpendicular to the plane we are going to consider. Let $x+u$, $y+v$, $z+w$ be the coordinates at the time t of a particle which if the solid were free from strain would be at (x, y, z). I will keep the same notation as in this article in the Encyclopedia Britannica.

The strain of the solid is the resultant of a simple longitudinal strain in the direction OX, numerically equal to $\frac{du}{dx}$, and two slips parallel to OY, OZ. The motion of one plane relatively to another may be thought of thus: Suppose these two books represent planes perpendicular to OX. The one part of the motion represented by u gives us a strain $= \frac{du}{dx}$. If for all values of x u is the same, the result will be only that the whole solid is pushed along. The strain, that is, the change of relative position of different parts of the solid is expressed so far as this part is concerned, by $\frac{du}{dx}$ in the regular notation. That is the simple longitudinal strain in the direction of OX. Think now what happens parallel to OY, viz, a slipping represented by these two books slipping past each other. The two other components then are shears corresponding to $\frac{dv}{dx}$, parallel to OY and $\frac{dw}{dx}$ parallel to OZ. The values of these shears, according to a general principle of evaluation of strains given in this paper are not to be reckoned by $\frac{dv}{dx}$, $\frac{dw}{dx}$, the simple shears. We take as unit shear the one in which the angle of distortion is $\frac{1}{\sqrt{2}}$, not 1, in the ordinary notation of a shear. A shear consisting in the change of shape of a square is normally represented by that angle in radians which is the diminution of one pair of right angles and the augmentation of the other pair. A simple distortion of strain, upon the principle set forth in this paper is reckoned in terms of another unit, a unit in which $\frac{1}{\sqrt{2}}$ would be the unit shear without anything more than infinitessimal shears admitted. Therefore $\sqrt{2}\,\frac{dv}{dx}$, $\sqrt{2}\,\frac{dw}{dx}$, are the numerical measures of the shears represented by $\frac{dv}{dx}$, $\frac{dw}{dx}$. I cannot go into the reasoning just now. You will find it distinctly set forth in Chapter X of this little article. I will just read Cor. 4 of that chapter:

"Cor. 4. A definite stress of some particular type chosen arbitrarily may be called unity; and then the numerical reckoning of all strains and stresses becomes perfectly definite." Ordinarily we choose as we please the unit. I have a reason for making all depend upon the unit which is chosen for one particular method of strain, which is fully set forth here. That is a proposition to be proved and made clear by illustrations. That being set forth, it remains for us to choose our unit. Following upon the proposition is this definition: "Def. A uniform pressure or tension in parallel lines, amounting in intensity to the unit of force per unit of area normal to it, will be called a stress of unit magnitude, and will be reckoned as positive when it is tension, and negative when pressure."

That definition being laid down, the previous proposition shows that we are no longer at liberty to represent a simple distortion by saying that it is the change of this right angle rather than some other change as for instance the elongation of the diagonal. I have two other sentences to read, so as to make my formula complete: "(4) A stress compounded of unit pressure in one direction and an equal tension in a direction at right angles to it, or which is the same thing, a stress compounded of two balancing couples of unit tan=

gential tensions in planes at angles of 45° to the direction of those forces, and at right angles to one another amounts in magnitude to $\sqrt{2}$." "(5) A strain compounded of a simple longitudinal extension x, and a simple longitudinal condensation of equal absolute value, in a direction perpendicular to it, is a strain of magnitude $x\sqrt{2}$; or, which is the same thing (if $\delta = 2x$), a simple distortion such that the relative motion of two planes at unit distances parallel to either of the planes bisecting the angles between the two planes mentioned above, is a motion δ parallel to themselves is a strain amounting in magnitude to $\frac{\delta}{\sqrt{2}}$."

Let us now consider the energy of the motion. Put $\frac{du}{dx} = \xi$, $\sqrt{2}\,\frac{dv}{dx} = \eta$, $\sqrt{2}\,\frac{dw}{dx} = \zeta$ .... (1) and let W denote the work per unit of bulk required to produce the distortion in question, irrespective of inertia. We have W a quadratic function of the three components of the strain, or, $W = \frac{1}{2}(A\xi^2 + B\eta^2 + C\zeta + 2D\eta\zeta + 2E\zeta\xi + 2F\xi\eta)$ .... (2) where A, B, C, D, E, F denote moduluses of elasticity of the solid. We shall consider a little more about obtaining these moduluses from the 18 moduluses of the solid. I merely say now however that these are the moduluses of elasticity (the definition of modulus of elasticity being "stress divided by strain") linearly obtained from any proper and sufficient data regarding the elasticity of the solid.

Let p, q, r denote the three components of the elastic traction per unit area of the wave front due to pulling these planes assunder and to their relative slipping parallel to OY and parallel to OZ. If the medium were isotropic then clearly the elastic traction resulting from these two planes would be a force opposing the traction parallel to OX and forces parallel to OY and OZ depending on the compressibility of the solid directly opposed to the slips in those directions. But generally, each one is involved in the other in the way that is expressed so conveniently by Green by the aid of the energy function, viz:

$$\left.\begin{aligned} p = \frac{dW}{d\xi} &= A\xi + F\eta + E\zeta, \\ q\sqrt{\tfrac{1}{2}} = \frac{dW}{d\eta} &= F\xi + B\eta + D\zeta, \\ r\sqrt{\tfrac{1}{2}} = \frac{dW}{d\zeta} &= E\xi + D\eta + C\zeta. \end{aligned}\right\} \cdots (3)$$

According to the notation here introduced, p, q, r being mere pulls, $p, q\sqrt{\frac{1}{2}}, r\sqrt{\frac{1}{2}}$ express the stress parallel to OX, OY, OZ respectively.

We want to find waves that will travel each with a given line of displacement. That is quite analogous to the problem of the fundamental modes of a vibrating body. Let us find, if we can, directions of displacements for which the return force will be in the direction of the displacement. The equations for that will be

$$\frac{dW}{d\xi} = \mathfrak{M}\xi, \quad \frac{dW}{d\eta} = \mathfrak{M}\eta, \quad \frac{dW}{d\zeta} = \mathfrak{M}\zeta \ \ldots (4),$$

where $\mathfrak{M}$ is constant. From these equations we can eliminate ξ, η, ζ forming the cubic equation in $\mathfrak{M}$:

$$(A-\mathfrak{M})(B-\mathfrak{M})(C-\mathfrak{M}) - (A-\mathfrak{M})D^2 - (B-\mathfrak{M})E^2 - (C-\mathfrak{M})F^2 + 2DEF = 0.$$

The three real roots obtained from this cubic will solve the problem. When the solid is strained in any one of these three ways, we have

$$p = \mathfrak{M}\frac{du}{dx}, \quad q = \mathfrak{M}\frac{dv}{dx}, \quad r = \mathfrak{M}\frac{dw}{dx} \cdots (5)$$

or the three components p, q, r will be proportional to the three displacements. The equations of motion, ρ being the density, are, of course, $\rho\frac{d^2u}{dt^2} = \frac{dp}{dx}$, $\rho\frac{d^2v}{dt^2} = \frac{dq}{dx}$, $\rho\frac{d^2w}{dt^2} = \frac{dr}{dx}$, which become on substituting the values of p, q, r from (5):

$$\rho\frac{d^2u}{dt^2} = \mathfrak{M}\frac{d^2u}{dx^2}, \quad \rho\frac{d^2v}{dt^2} = \mathfrak{M}\frac{d^2v}{dx^2}, \quad \rho\frac{d^2w}{dt^2} = \mathfrak{M}\frac{d^2w}{dx^2}.$$

These by equations (4) and (1) give the formulae

$$\left.\begin{aligned} Au + (Fv + Ew)\sqrt{2} &= \mathfrak{M}u \\ Fu + (Bv + Dw)\sqrt{2} &= \mathfrak{M}v\sqrt{2} \\ Eu + (Dv + Cw)\sqrt{2} &= \mathfrak{M}w\sqrt{2} \end{aligned}\right\} (6)$$

Let $\mathfrak{M}_1$, $\mathfrak{M}_2$, $\mathfrak{M}_3$ be the roots of the determinantal cubic and b_1, c_1, b_2, c_2, b_3, c_3 the corresponding values of the ratios $\frac{v}{u}$, $\frac{w}{u}$, derived from (E). Observe that $u = u_1$, $v = b_1 u_1$, $w = c_1 u_1$ is a solution, where $u_1 = f_1\left(x + t\sqrt{\frac{\mathfrak{M}_1}{\rho}}\right) + F_1\left(x + t\sqrt{\frac{\mathfrak{M}_1}{\rho}}\right)$, and the thing is done. That is the full investigation for one of the three waves. The velocity of preparation is $\sqrt{\frac{\mathfrak{M}_1}{\rho}}$. For the other two waves you can write down similar expressions corresponding to the second and third roots, $\mathfrak{M}_2$, $\mathfrak{M}_3$.

Lecture XII.

We will look a little more at this wave problem. I do not know that I should have troubled you with going through a process like this, because you will find it easier to read it in the book. The conclusion is, that if you choose arbitrarily, in any position whatever relatively to the elastic solid, a set of parallel planes for wave fronts, there are three directions at right angles to one another (each oblique to the set of planes) which fulfil this important condition, that the elastic force is in the direction of the displacement: and the equations we put down express the wave motion. Each of the three waves will be a wave in which the oscillation of the matter in its front is as I am performing it now, i.e., an oscillation to and fro in a line oblique to the plane of the wave front. You will find the three waves corresponding to the three roots of the determinantal cubic are in directions at right angles to one another and in general oblique to the plane of the wave.

Green deals with the problem in a peculiar way. He expresses the conditions by means of three equations among the coefficients that in two of these three waves possible to an elastic solid, the displacement is exactly in the wave front, giving two waves at right angles to each other in the wave front and the third wave in the direction perpendicular to the wave front. The result of those three relations that Green finds among the coefficients – Green does not say anything of this, but we will think of it a little – will be that in considering a disturbance from a source we have a wave of distortion proceeding outwards corresponding to, but not in all cases identical with, that of Fresnel – made identical with that of Fresnel by some other supposition which I shall not speak of now. You see, if you work out the mathematical array of figures you might put down the equations in all their generality. I do not think this has ever been done. But first take the process we have gone through with, not for a plane perpendicular to OX as we did it, but for a plane oblique to the three axes, and you get three velocities for waves perpendicular to any chosen direction. Then by taking the envelope of these with the proper mathematical conditions, which you can put down in a few moments, and you get a wave surface which will differ from anything that has been thought of before, so far as I know in the theory of elastic solids — a wave surface in which there will be three sheets corresponding to each radius vector instead of only two, as in Fresnel's wave surface; and so far as

I have investigated, each part of that wave surface will involve both condensation and distortion. I have satisfied myself that this is so. It is a geometrical exercise of no contemptible character to work out this wave surface. Green's three relations cause this wave surface to split up into an ellipsoidal wave surface for a condensational wave and a wave surface like Fresnel's for a distortional wave. I say "ellipsoidal"; so far as I remember, Green does not mention it at all. That is an exceedingly interesting result, and Green's three relations that give rise to it are exceedingly interesting relations.

Look back at the formulae which we had in an isotropic solid. We have always a perfect distinction between the two waves; that is we can have a purely distortional wave spreading out with its velocity, and a purely condensational wave with its velocity. You will remember the condensational wave proceeding symmetrically from a center, for which $\Phi = \frac{1}{r} \sin \frac{2\pi}{\lambda}(r - vt)$ was the displacement potential, or the condensational wave for which any differential coefficient of Φ whatever, $\frac{d^{i+j+k}\Phi}{dx^i dy^j dz^k}$, is the displacement potential, all of which proceed with the same velocity $v = \sqrt{\frac{k + \frac{4}{3}n}{\rho}}$. Take $n = 0$ and you have what is dealt with in Lord Rayleigh's work on sound, in which he takes not only the distance terms as I did, but terms that express the motion at distances from the center moderate in comparison with the wave length. For an isotropic solid we can have all those waves, and again simultaneously with them, we can have some of our solutions for waves of distortion. Take, if you like our solution $\frac{4\pi^2}{\lambda^2}\Phi + \frac{d^2\Phi}{dx^2}, \frac{d}{dy}\frac{d\Phi}{dx}, \frac{d}{dz}\frac{d\Phi}{dx}$.

is a fundamental solution, from which by differentiation you can obtain other solutions. I am going to correct something that I said as to want of interest. This solution is more interesting than I thought it would be.

In the isotropic solid the independency of the two sorts of waves comes naturally. The condensational wave goes at one speed, the distortional goes at another and you may proceed with either as if the other were not there. A central disturbance in an isotropic solid will cause both sorts of waves to proceed outwards in the manner of an earthquake eruption. I do hope that before another earthquake will do as the last one did, there will be means of observing earthquake disturbances. There is a great deal to investigate in that subject. I do think it will be worth while at stations for the observation of scientific meteorology to have self recording apparatus to show the three components of apparent gravity at every instant. Beyond a doubt, if there had been records taken in this way, by instruments not too sensitive during the last earthquake, we should have had evidence of two waves through the earth, a distortional and a condensational wave.

What goes on in the isotropic solid occurs similarly in a solid which is not isotropic, but which fulfils Green's conditions. I may tell you that these conditions are first a set of conditions of symmetry, and secondly, three equations. I have not looked into the thing to see whether, without other conditions than that of the condensational wave having its own wave surface and set of velocities, a distortional wave may be capable of propagation without condensation at all. The condition for that alone, has not so far as I know been investigated. Green only gives the condition for that after having introduced certain conditions of symmetry; and I do not know whether it would come to the same result or not if he had introduced it before giving the conditions of symmetry. That is a very interesting subject and I shall attempt to work it out. I think you will find it worth while to work out the wave surface that I gave and then Green's interesting condition, what must be the condition that two out of the three waves whose fronts are parallel to a given plane shall be purely distortional. It is obvious that if this condition is fulfilled, without any other consideration of symmetry, you will have an unsymmetrical quasi-Fresnel wave surface for purely distortional waves, which, when made symmetrical with reference to OX, OY, OZ, by a series of relations, will become more like Fresnel's, but which certainly requires something more of an assumption than that to make it agree with Fresnel's.

That is a fine subject for investigation, and I am sorry I can not throw more light upon it. There is nothing more in it than would take a mathematician half a day to work out, and it would be worth doing.

But if the war is to be directed to fighting down the difficulties to the undulatory theory of light it is not the slightest use for us in solving our difficulties to have a medium which kindly permits distortional waves to be propagated through it, though it is aeolotropic. It is not enough to know that though the medium be aeolotropic it can let purely distortional waves through it, and that two out of the three waves will be distortional. What we want is a medium which, when light is refracted and reflected, will under all circumstances give rise to distortional waves alone. Green's mediums would fail in this respect when waves of light come to a surface of separation between two such mediums. All that Green secures is that there can be an outward distortional wave; he does not secure that there shall not be a condensational wave. There would be condensational waves from the source. The electric light, etc., would produce condensational waves, whether it was in an aeolotropic or isotropic medium so far as Green's conditions here spoken of go. Interesting as they are, they do not help in the slightest degree towards explaining double refraction in such a medium. What we want is a medium resisting the condensational waves; a medium with an infinite or practically infinite bulk modulus – so great that there can never be more than an amount of energy that has not been discovered by observation, developed in the shape of condensational waves. – I believe that is a correct sentence although it is complicated.

As an essential in every reflection and refraction there may be a little loss of energy from the want of perfect polish in the surface, but as a rule we have no loss of light in reflection and refraction. There perhaps is some and we have not discovered it. The medium that gives us the luminiferous vibrations must be such that if there is any part of the energy of the wave expended in condensational waves after refraction and reflection, the amount of it must be so small that it has not been discovered. Numerical observations have been made with great accuracy, in which for example, Fresnel's formula for the ratio of the incident and reflected light $\left(\frac{\mu-1}{\mu+1}\right)^2$ is verified within closer than one per cent, I think. Still a half per cent or a tenth per cent of the energy may be converted into condensational waves, for all we know. But if any per-centage to speak of were converted into condensational waves, there would be a great deal of energy in condensational waves going about through space, and there would be a new force (to take an absurd mode of speaking of these things) that we know nothing of. There would be some tremenduous action all through the universe produced by the energy of condensational waves if the energy of condensational waves were one-tenth, or one-hundredth, or even say one-thousandth per cent. of the energy of the distortional waves. I believe that if in all instances of reflection or refraction of light at any surface or in case of violent action in the source, there are condensational waves produced with anything like a thousandth or a ten-thousandth of the energy of light, we should have some prodigious effect, but which might, perhaps have to be discovered by some other senses than we have. The want of indication of any such actions is sufficient to prove that if there are any in nature, they must be exceedingly small. But that there are such waves I believe, I am. I believe that the velocity of propagation of electro-static force is the unknown condensational velocity that we are speaking of.

I say "believe" here in a somewhat modified manner. I do not mean that I believe this as a matter of religious faith, but rather as a matter of strong scientific probability. If this is true of propagation of electro-static force, it is perfectly true that there is exceedingly little energy in the waves corresponding to the propagation of an electro-static force. That is going beyond our tether however, of Molecular Dynamics. What I proposed in, the introductory statement with reference to these Lectures was to chiefly bring what principles and results of the science of molecular dynamics I could enter upon to bear upon the wave theory of light. We are sticking closely to that for the present, and we may say that we have nothing to do with condensational waves. Our medium is to be incompressible, and instead of Green's three conditions, we have one condition of incompressibility. It is obvious that one equation of incompressibility suffices to prevent the possibility of a wave of condensation at all and reduce our wave surface to a surface with two sheets, like the Fresnel surface. But before passing away from that beautiful dynamical speculation (or example of possibility I should perhaps call it) of Green's, if we think of what the condensational wave must be in an aeolotropic solid fulfilling Green's condition that it can have purely distortional waves proceeding in all directions — the condition that two of the three waves we investigated three-quarters of an hour ago shall be purely distortional — I think we shall find also condensational waves, and that the wave surfaces for them will be a set of concentric ellipsoids. It will be a single sheeted surface, that is certain, because you have only one velocity corresponding to each tangent plane at the wave surface.

I shall now leave this subject for the present. We shall come back upon it again, perhaps, and look a little more into the question of moduluses of elasticity. We shall work up from an isotropic solid to the most general solid; and we shall work down from the most general solid to an isotropic solid. We shall take first the most general value for the compressibility; we shall then come to this subject again of assuming incompressibility. We shall then begin with the most general solid possible, and see what conditions we must impose to make it as symmetrical as is necessary for the Fresnel wave surface. The molecular problem will prepare your way a good deal for this.

That puts me in mind of a correction I have to make with respect to the interest attached to this solution for distortion in an isotropic solid $\left\{\frac{4\pi^2}{\lambda^2}\varphi + \frac{d^2\varphi}{dx^2}, \frac{d^2\varphi}{dxdy}, \frac{d^2\varphi}{dxdz}\right\}$ I said it was not interesting because it could not express a natural sequence of light waves. I said that to express a natural sequence of light waves we must have two bodies moving in opposite directions, so that the center of gravity may not move. I quite forgot the supposition of our shell, which does the very thing we want. Instead of passing to a higher order of differentiation, so to speak, for the most probable natural sequence of waves of light, consisting of waves of greater nodal subdivision, (having a nodal circle at the equator as well as nodal points in the axis of x) I see now that this very thing is the most probable. By "probable" I mean, certainly the most frequent. I look upon it as a reality that there are particles moving; and it seems to me certain that those particles are soft and that they must have enormous mass compared with the luminiferous ether.

I had intended to prepare something about the mass of the luminiferous ether. I have not had time to take it up, but certainly shall do so before we have done with the subject. We shall go into the question of the density of the luminiferous ether, giving superior and inferior limits. We shall also consider what fraction of a gramme may be in one of these molecules and show what an enormously smaller fraction of a gramme we may suppose it to displace in the luminiferous ether. We shall try to get into the notion of this, that the molecule must be soft and that that there must be an enormous mass in its interior. Its outer part feels and touches the luminiferous ether and the luminiferous ether feels, it may be, comparatively slight to it. It is a very curious supposition to make, of a molecular cavity lined with a massless rigid spherical shell; but that something exists in the luminiferous ether and acts upon it in the manner that is faultily illustrated by our mechanical model, I absolutely believe. I have no more doubt that something of the kind is true, than I have of my own existence.

Just think of the effect of a shock consisting say of a collision between that and another molecule. Instead of its being broken into bits, let us suppose a case around it. It will bound away, vibrating. Just imagine that the central nucleus goes in one direction while the shell is going in the other, and there will be a molecule with two parts going in opposite directions but different from what I thought of the other day in that one part is inside the other. The ether gets its motion from the outside part. Therefore I say that the most fundamental supposition we can make with reference to the origin of a sequence of waves of light is that illustrated by a globe vibrating to and fro in a straight line.

We have already investigated the solution corresponding to that. Take spherical waves; no vibrations for points in one certain diameter of the sphere; maximum vibrations in all points of the equatorial plane of that diameter and perpendicular to that plane; for all points in the quadrant of an arc of the spherical surface extending from axis to equator, vibrations in the plane of and tangent to the arc of magnitude proportional to the cosine of the latitude or angular distance from the equator and of intensity proportional to the square of the cosine of the latitude. Then let the amplitude vary inversely as the distance from the center; and the intensity inversely as the square of the distance from the center, and you have a correct word-painting of the very simplest and most frequent sequence of vibrations constituting light.

Let us return to the consideration of the dynamics of refraction, absorption, anomalous dispersion, and so on. We have the square of the refractive index, $\frac{n\tau^2}{\rho\lambda^2} = \mu^2 = 1 + \frac{c_1}{\rho}\left\{q_1\kappa_1^2 - (1-q_1)\tau^2 + q_1\kappa_1^2\left(\frac{\kappa_1^2}{\tau^2} + \frac{\kappa_1^4}{\tau^4} + \cdots\right) - q_2\frac{\tau^4}{\kappa_2^2-\tau^2} - \cdots\right\}$ We are not going to think at present of what the values of q_1, q_2 .. may be, except that q_1 is less than unity. Our example will illustrate that, but I think we can see it without an illustration. For explaining ordinary refraction we probably have more than enough in the term $q_1\kappa_1^2$, and it will be convenient to suppose all the other terms small in comparison with this. When τ is small in comparison with $\kappa_2, \kappa_3, \ldots$ the terms in q_2, q_3 are negative, but I believe we can suppose $q_2, q_3, \ldots$ all exceedingly small. Remember that $q_1, q_2, \ldots$ are constants, independent of the period τ, as well as $\kappa_1, \kappa_2, \ldots$ which are the periods in which the system would vibrate if we hold the handle P at rest, i.e. the case of $\xi = 0$ in our equations.

I want to see how we can vary τ without coming to trouble. As we increase τ, the negative term becomes larger and larger, and if we increase it enough it will make $\mu^2 = 1$; increase it still more and it will make $\mu^2 = 0$ and if we increase it still more, it will make μ^2 negative. Let us put this in its other form. This form is only suitable to show its availability for modifying Cauchy's formula so as to give correct refractions. I hope we may have a little work done upon it sometime or other in the way of seeing whether these terms will suffice for actually obtaining refractive indices. I believe Lommel has done something of this kind. I know Ketteler in 1871, had a formula quite like this, but if I remember right, his term was positive here instead of essentially negative, as I have it. It seems probable that we should be able to explain refraction through a somewhat wide range from what is here written. You are doubtless more familiar with the formula in which λ appears, but remember that λ is proportional to the period, so that this formula is simply $A - B\lambda^2 + C\lambda^{-2} + D\lambda^4$, which falls back upon Cauchy's formula $A + B\lambda^{-2} + C\lambda^{-4}$, except through a very wide range, or to meet the critical case.

I am sorry to leave it but we must. I have data from Langley for the refrangibility of different lights passing through rock salt, down to about three or four times the wave length of sodium light by actual observation, if I remember right, and by very probable inference from the curve obtained down to 7 times the wave length of sodium light. I received this only a day or two ago, so that I have not attempted to make a comparison with a formula of this kind.

I am going to ask you to look at the critical case, and for this purpose replace this form by the simpler one:

$$\mu^2 = 1 + \frac{C_1 T^2}{\rho}\left(-1 + \frac{q_1 T^2}{T^2 - \kappa_1^2} - \frac{q_2 T^2}{\kappa_2^2 - T^2} - \ldots\right)$$

in which T is greater than κ_1 and less than κ_2. When T is considerably larger than κ_1, but small in comparison with κ_2, we have ordinary refraction in a transparent body, without absorption bands, or anything of the sort. This must occur through a considerable range of values of T in the cases of glass, rock salt, etc. As T decreases we have an augmenting refractive index for ordinary normal refraction. As T approaches κ_1, μ^2 approaches infinity, and you get greater and greater refraction, until you pass through κ_1. When κ_1 exceeds T what will the result be? μ^2 will become negative. What is the meaning of the square of the refractive index negative? Answer, waves cannot be propagated. Think of the proposition, waves cannot be propagated at all. That is clearly an absorption band.

I object to the invoking of viscous terms to get quit of the energy; for how shall we prevent them from taking away all our energy when we do not want it taken away? We can spare exceedingly little energy in the transmission of light through distilled water, if it may be propagated through 150 feet as I believe it is. Sea water is supposed to be more transparent than most bodies. It is by no means black darkness down 20 fathoms in any sea. There are about 500 000 wave lengths of sodium light in a foot of water. In 100 feet there would be 50,000,000 wave lengths. We can spare very little energy, then, in water, if we are to think of light being propagated through 50 million wave lengths before it is absorbed. If we let in viscous terms in a way that will do anything at all for us in answering the question, what becomes of the energy at the critical points, for the wave lengths that are actually absorbed, it will run away with our energy where we do not want it to. Besides that, it is throwing up the sponge in respect to the dynamical question, and confessing that we have to introduce a new force instead of dealing solely with dynamical ones. As a subordinate theory in abstract hydrodynamics, it is exceedingly interesting to introduce viscous terms; but not in molecular dynamics. We must think of what becomes of the energy. Helmholtz understands, as I said before, that the consumption of energy by the viscous terms means its conversion into heat. But I want the same vibrating molecule which gives us the ordinary laws of refraction, which gives us the anomalous dispersion at the critical points, to take up the energy also and give it out in the proper way. That is what we have been doing thus far. And I want to look at a set of vibrating particles, and see what may be obtained from them. Of course we can do far more by calculation than we can find out in that sort of way, but still, it will help us a little.

It is perfectly clear that we have a broad absorption band throughout the range of values of T smaller than κ, which gives a negative value to μ^2. The light will first appear beyond that absorption band with very small refrangibility — exceedingly small. We have in the neighborhood of the critical point exactly that kind of inversion with anomalous dispersion, in which we have less refraction, or greater velocity of propagation, for light of periods less than a certain limit, κ, say, and greater refractive index or less velocity of propagation for light above that period.

That is merely an indication of the fact of anomalous dispersion; it is hardly worth while to look into the details just now. That is doubtless familiar to many of you, who have read Helmholtz's paper on anomalous dispersion. The subject was worn threadbare before I knew it was discovered, in fact. Stokes originated the idea of periods of absorption corresponding to periods of molecular vibration; but there was no hint of explaining refraction in this way or anomalous dispersion. So far as I know the first word on the reaction of these particles upon the luminiferous ether is Sellmeier's; and it is now perhaps, a matter of general knowledge and I should rather apologize for taking up the time in speaking of it than to say that I am bringing anything new before you.

We shall try to see something more of the effect of light propagated through a medium of.

a period exactly equal to x_1. I believe each sequence of vibrations will throw in a little energy which will spread out among the different possible vibrations of the molecule. The combination of the sequences, forming what we call continuous light, is not a continuous phenomenon at all. I believe that the first effect when light begins will be, each sequence of waves of the exact period throws in some energy into the molecule. That goes on until, somewhere or other, the molecule gets uneasy. It takes in an enormous quantity of energy before it begins to get particularly uneasy. It then moves about, and begins to collide with its neighbors perhaps, and will therefore give you heat in the gas, (if it be a gaseous molecule). It goes on colliding with the other molecules, and in that way imparting its energy to them. The energy will be simply carried away by convection if you please, or a part of it perhaps. Each molecule set to vibrating in that way becomes a source of light, and so we may explain the radiation of heat from the molecule after it has been got into the molecule by sequences of waves of light. I believe we can so explain the augmented pressure of a gas, due to the absorption of heat in it.

We may consider, however, that the chiefest vibration of the molecule is that in which the nucleus goes in one direction and the shell in the opposite direction, but with a great amount of energy in the interior vibrations and very little in the shell so that the shell may go on giving out phosphorescent energy for two or three hours or days, simply vibrating forever, except in so far as the energy is drawn off and allowed to give motion to other bodies.

I see no difficulty in answering several of the fundamental riddles of this subject by the reaction of this assumed particle in the luminiferous ether; but there is difficulty about double refraction, and I see no solution whatever of that riddle as yet.

Lecture XIII.

Prof. Morley has solved the problem that I proposed for some of the fundamental periods, and you may be interested in knowing the result. He finds roots $\frac{1}{\tau^2}$ = 3.46, 1.005, .298, .087, each root not being very different from three times the preceding one. A tracing of the curve, you will understand, involves a set of asymptotes. The curve in general for any such case must be something like this:

, the curve going in this direction (arrow) from positive to negative. At the end $\frac{1}{\tau^2} = \infty$. I will not go into that any further just now. I just wanted to call your attention to what Prof. Morley has done upon the example that I gave to the arithmetical laboratory. I think it would be worth while, also, to work out the energy ratios. In selecting this example, I chose the case for which the work would of necessity be highly convergent. But I chose it primarily, however, because it is something like the kind of thing that presents itself in the true molecule:— A soft elastic body consisting of a finite number of discontinuous masses elastically connected, (with enormous masses in the central parts, that seems certain) imbedded ether and acted on by the ether in virtue of an elastic connection which, if this molecule were rigid and imbedded in the ether simply like a rigid mass imbedded in jelly, must consist of elastic bonds analogous to springs.

I think you will be interested in looking at this

model which, by the kindness of Prof. Rowland, I am now able to show you. It is made on a plan according to which I made a wave machine which has been used for many years in my classes, and finally modified in preparations for a lecture given to the Royal Institution about two years ago on The Size of Atoms. I think those who are interested in the

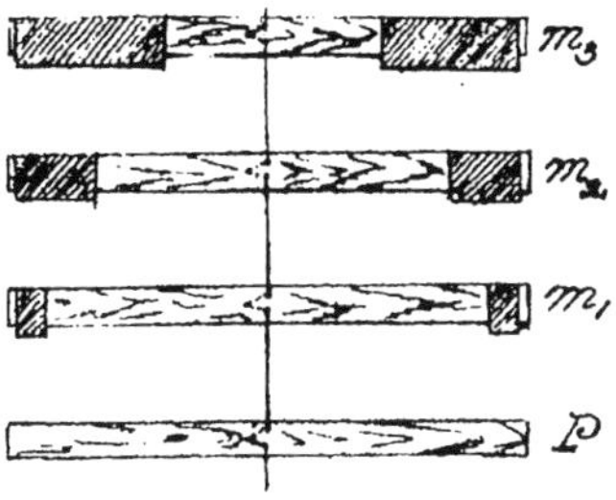

illustration of dynamical problems will find this a very nice and convenient method. If you will look at it, you will see how the thing is done: Piano forte wire bent around three pins in the way you see here , supporting each bar. Those pins are slanted in such a way as to cause the wire to press in close to the bar so as to hold it quite firm. The wood is slightly cut away to prevent the wire from touching it so that there may be no impairment of elasticity due to slip of steel on wood. The wire used is fine steel piano-forte wire; that is the most elastic substance available, and it seems to me indeed by far the most elastic of all the materials known to us.

Prof. Rowland is going to have another machine made, which I think you will be pleased with — a continuous wave machine. This is not a wave machine, but a machine for illustrating the vibrations of several elastically connected particles. The connecting springs are represented by the torsional spring in the three portions of connecting wire and the fourth portion by which the upper mass is hung. In this case gravity contributes nothing to the effect except to stretch the wire. You will understand that these upper masses correspond to m_1, m_2, m_3. In all we have four masses here. I will just apply a moving force to this lower mass P. To realize the circumstances of our case more fully, we should have a spring connected with a vibrator to pull P with, and perhaps we may get that up before the next lecture. I shall attempt no more at present than to cause this first particle to move to and fro in a period which is perceptibly shorter than the shortest of the three periods. The result is scarcely sensible motion of the others. I do not know that there would be any sensible motion at all if I had observed to keep the greatest range of this lowest particle to its original position on the two sides of its mean position.

The first part of our lecture this evening I propose to be a continuation of our conference regarding aeolotropy. The second part will be molecular dynamics. I propose to look at this question a little, but want to look very particularly to some of the points connected with the conceivable circumstances by which we can account for not merely regular refraction but anomalous dispersion and both the absorption that we have in liquids and very opaque bodies and such absorption as is demonstrated by the existing fine lines of the solar spectrum which are now shown more splendidly than ever by Prof. Rowland's gratings.

I shall speak now of aeolotropy. The equations by which Green realized the condition that two of the three waves having front parallel to one plane shall be distortional is equivalent to a very easily understood condition that I may illustrate first restricting bodies more nearly isotropic than those that we are considering in the more general problem.

I am reminded by a lady that I have said two or three times dilatational instead of distortional* and I have just said it again. There seems to be a law by which I say dilatational when I mean distortional.

* [This has been corrected wherever I have noticed it. H.]

Another little point with respect to yesterday's work: if you have taken the trouble to make notes, you had better cancel the $\sqrt{2}$ wherever it occurs, and let the unit tangential stress be the ordinary unit as set forth in Thomson and Tait for example, and the unit of distortion a simple shear. There is good reason for the $\sqrt{2}$, but it is a part of the theory that we are not concerned with at all, and for a special problem like that, it is better to introduce special notation. This special notation is in point of fact the more general notation.

That problem is similar to another of the very greatest simplicity which is the well known problem of the displacement of a particle subject to forces acting upon it in different directions from fixed centres. An infinitesimal displacement in any direction being considered the question is, when is the return force in the direction of the displacement. As we know, there are three directions at right angles to one another in which the return force is in the direction of the displacement. The sole difference between that very trite problem and that which I went through yesterday is that in the latter case the question is put with reference to a whole infinite plane in an infinite homogeneous solid which is displaced in any direction. Considering force per unit of area, we have the same question, when is the return force in the direction of the displacement and the answer is, there are three directions at right angles to one another in which the return force is in the direction of the displacement. Those three directions are generally oblique to the plane: but Green found the conditions under which one will be perpendicular to the plane, and the other two in the plane.

I shall now enter upon the subject more practically in respect to the application to the wave theory of light, and that is, to introduce right away at the beginning the condition of incompressibility. Take first the well known equations of motions for an isotropic solid and express in them the condition that the body is incompressible. The equations are: $\rho \frac{d^2\xi}{dt^2} = (k+\frac{1}{3}n)\frac{d\delta}{dx} + n\nabla^2\xi$, etc.

I have another name from Prof. Ball for ∇, which is atled, or delta spelt backwards. Shall it be nabla, atled, or Laplacian? Laplacian, if you like.

Suppose now the resistence to compression is infinite, which means, make $k=\infty$ at the same time that we have $\delta=0$. What then is to become of the first term of the second members of these equations? We simply take $(k+\frac{1}{3}n)\delta = p$, and write the second member $\frac{dp}{dx} + n\nabla^2\xi$. This requires no hypothesis whatever. We may now take $k=\infty$, $\delta=0$, without interfering with the form of our equations. These equations, without any condition whatever as to ξ, η, ζ, with the condition $p=(k+\frac{1}{3}n)\delta$ are the equations necessary and sufficient for the problem. On the other hand, if $k=\infty$, the condition that that involves is $\frac{d\xi}{dx}+\frac{d\eta}{dy}+\frac{d\zeta}{dz}=0$ which gives four equations in all for the four unknown quantities ξ, η, ζ, p.

Precisely the same thing may be done for a solid with 21 independent coefficients. We will have this equation again for an aeolotropic body, $\delta=0$, and a corresponding equality to infinity. I am not going to introduce any of these formulae at present. In the meantime, I tell you a principle that is obvious. In order to introduce the condition $\frac{d\xi}{dx}+\frac{d\eta}{dy}+\frac{d\zeta}{dz}=0$ into our general equation of energy with its 21 coefficients which involves a quadratic expression in terms of the six quantities that we have denoted by e, f, g, a, b, c, we must modify the quadratic into a form in which we have $(e+f+g)^2$ into a coefficient. That coefficient equated to infinity, and $e+f+g=0$, leave us the general equations of equilibrium of an elastic solid with one fewer out of the 21 independent coefficients in virtue of this relation of incompressibility.

I want to call your attention to the kind of deviation from isotropy which is annuled by Green's equations among the coefficients which express that two out of the three waves shall be purely distortional. The next thing to an isotropic body is one possessing what Rankine calls cyboid symmetry.

Rankine marks an era in phylology, and scientific nomenclature. In England, and I believe in America also, there has been a classical reaction or reformation according to which, instead of taking all our Greek words through the French changing k into c, υ into y, and several other variations that I do not remember, we spell in English, and pronounce Greek words, and even some Latin words more nearly according to what we may imagine to be the actual usage of the ancients; We cannot however get over Kuros instead of Cyrus, Kikero instead of Cicero, in the present generation; we have not swallowed the thing altogether, yet. Rankine is a curious specimen of the very last of the French classical style. Rankine was the last writer to speak of cinematics instead of kinematics. Cyboid is a very good word, but I do not know that there is any need of introducing it instead Cubic. Cubic is an exception according to the regular analogy in that u is not changed into y; it should be cybic. I suppose κυβας to be the Greek word, because cyboid obviously means cubic, and it is taken from the Greek in Rankine's manner.

Rankine gives the equations that will leave cubic symmetry. He afterwards makes the very opposite remark that Sir David Brewster discovered that kind of variation from isotropy in analcime. I only came to this in Rankine two or three days ago. But I remember going through the same thing myself not long ago and I said to Stokes — I always consulted my great authority Stokes whenever I got a chance — "Surely there may be such a thing found to exemplify this kind of asymmetry; would it not be likely to be found in crystals of the cubic class?" Stokes — he knew almost everything — instantly said "O, Sir David Brewster thought he had found it in cubic crystals, but there was an explanation that it seemed to be owing to the effect of the cleavage planes, or the separation of the crystal into several crystalline laminæ" — I do not remember what it was, but he distinctly denied that Brewster's experiment showed a true instance of cubic asymmetry. He pointed out that an exceedingly slight deviation from cubic isotropy would show very markedly on elementary phenomena of light, which might be very readily tested by means of ordinary optical instruments, and that the fact that nothing of the kind has been discovered is absolute evidence that the deviation, if there is any, from isotropy in a crystal of the cubic class, is exceedingly small in comparison with the deviation from isotropy presented by ordinary double refracting crystals. Deviation from cubic isotropy is the same thing as the conceivable cubic deviation from isotropy.

As a matter of fact, deviation from square isotropy is found in a pocket handkerchief or piece of square cloth, supposing the warp and woof to be accurately similar a supposition that does not hold of ordinary cloth. Take wire cloth carefully made in squares and that will be symmetrical and equal in its moduluses with reference to two axes at right angles to one another. There will be a vast difference according as you pull out one side and compress the other, or pull out one diagonal and compress the other. Take the extreme case of a cloth woven up with inextensible frictionless threads, and there is a kind of absolute resistence to distortion in two directions at right angles to one another, and no resistence at all to distortion of a certain kind that is presented in changing its square shape. That is to say, a frame work of this kind 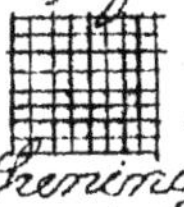 has no resistence to shearing distortion; but it has infinite resistence to the distortion produced by lengthening one diagonal and shortening the other. Just imagine a square cut out of this pattern with sides parallel to the diagonals, making a pattern of this sort  There is a body that has infinite resistence to shearing, and zero resistence to pulling out in this direction (along the diagonal). That is not altogether a trivial illustration. Surgeons make use of it in their bandages. A person not familiar with the theory of elastic solids might cut a strip lengthwise with the thread; but cut it obliquely and you have that conveniently pliable character that allows it to serve the purpose of a bandage.

Imagine an elastic solid made up in that kind of way, with that kind of deviation from isotropy and we have clearly two different rigidities for different distortions in the same plane. I remember that Rankine, in one of his early papers proved that to be impossibile. He proved a proposition to the effect that the rigidity was the same for all distortions in the same plane. That perhaps was founded on some special supposition as to arrangement of molecules and may be true for the particular arrangement. Rankine made too short work of the elastic solid in his first paper. He afterwards took it up very much on the same foundation that Green did with 21 coefficients, but he uses the old proposition that rigidity is the same for all distortions in the same plane.

I will go no further into that just now than to say that if without introducing the condition of incompressibility at all, you introduce the condition that there is equality of rigidities for the two principal modes of distortion in each plane — Perhaps we shall be able to face the problem in the next lecture of introducing the relations among the 21 moduluses which are sufficient to do away with all obliquities with reference to the rectangular axes. I shall put down the figures before you somehow or other, before we have ended. But you can do this in a moment — equate to zero enough of the 21 coefficients to fulfil these conditions, that if you compress the body by uniform forces parallel to OX or OY or OZ, it will remain rectangular, and that if you produce a shear in one coordinate plane it does not produce obliquity in any other, and so on, doing away with all that is necessary in order to annul obliquity. There will remain a certain number of coefficients — nine I think. Now put in your condition that in this plane the rigidity due to a shear parallel to the sides is equal to the rigidity due to a shear in a portion cut out with its sides at angles of 45° to the sides of this. There will be three equations. These equations are identical with the three equations that Green gives to express his condition as to the waves. That is really very interesting and instructive, although it does not do much for light.

I must read to you some of Rankine's fine words that he has introduced into science in his work on the elasticity of solids. That is really the first place I know of except in Green in which this thing has been gone into in a satisfactory way. It is not really satisfactory in Rankine except in the way in which he carries out the whole subject, the algebra of it and the determinants and matrices that he goes into so very nicely and, what I want to call attention to, his names. I do not know whether Prof. Sylvester ever looked at these names. I think he would be rather pleased with them. "Thlipsinomic transformations" "Umbral surfaces" and so on. Any one who will learn the meaning of all these words will obtain a large mass of knowledge with respect to an elastic solid. The words of "strain and stress" are due Rankine; "potential" energy also. Hear the grand words "Thlipsinomic, Tasinomic, Platythliptic, Euthytatic, Metatatic Heterotatic, Plagiotatic, Orthotatic, Pantatic, Cybotatic, Goniothliptic, Euthythliptic, &c."

You may now understand what cuboid asymmetry is, or as I prefer to call it, cuboid aeolotropy. Rankine had not the word aeotropy; that came in later. Cuboid or cubic aeolotropy is the kind of aeolotropy exhibited by a cube grating, a basket woven solid with uniform cubic baskets. There is a thing that would be isotropic, except for that difference of rigidity for the two principle distortions in each one of the planes of symmetry. What I am going to do further is to point out that if we take, first of all, the condition of infinite resistence to compression, secondly, introduce the conditions necessary for symmetry, then after that annul the difference of rigidities for the principle distortions in each of the three principal planes we shall find ourselves landed in an elastic solid with three principle moduluses which will give us a wave surface identical with Fresnel's, except that the order of procedure is different. The direction in this surface which corresponds to the direction of vibration in Fresnel's surface is a line perpendicular to the plane through the line of displacement and the perpendicular to the wave front, I believe; but it is possibly the plane through line of displacements and the center of the wave surface I will read it out of Green; but Green really never introduced the condition of incompressibility at all. Here it is at the bottom of page 304 of Green's collected papers, "We thus see that if we conceive a section made in the ellipsoid to which the equation (10) belongs, by a plane passing through its center and parallel to the wave's front, this section, when turned 90 degrees in its own plane, will coincide with a similar section of the ellipsoid to which the equation (8) belongs, and which gives the directions of the disturbance that will cause a plane wave to propagate itself without subdivision, and the velocity of propagation, parallel to its own front. The change of position here made in the elliptical section is evidently equivalent to supposing the actual disturbances of the ethereal particles to be parallel to the plane usually denominated as the <u>plane of polarization</u>

Thus, in the wave surface corresponding to Green's elastic solid, draw a plane perpendicular to the wave front through the direction of displacement. The line perpendicular to that plane is the direction of displacement in Fresnel's case.

I gave you one solution of the problem of passing a cord around the eight vertices and twelve edges of a parallelopiped. It is obvious that it cannot be done by passing the cord only once along each side. To make the figure incompressible, we may suppose the cord to be perfectly flexible and inextensible. Instead of supposing the cord inextensible, we can have an elastic portion in the middle part of the cord along each side. You can thus introduce what is equivalent to three disposables in the longitudinal rigidities of the portions of the cord in question. We may dispense with the idea of a flexible body if wherever the cord changes direction we put in a bell handle, which is a mechanical principle, instead of passing the cord through a ring.

I am afraid this problem of the molecules in the elastic solid presents enormous difficulties to us. I feel that we have the utmost confidence that we can make a model that will fulfil any stated condition whatever, as to absorption, and so on. The mathematical working of it out is difficult. I am not going to solve all these problems in five minutes but what I can do in five minutes is to show that we are quite out of our depth after all, in the thing we have been invoking. Consider the uniform isotropic elastic solid in which this molecule is imbedded. We must consider the distance from one of these imbedded shells to another to be great in comparison with the diameter of the shell and small in comparison with the wave length. If the effect is anything near sufficient to give us change of velocity through the range of 1 to 1.5, we cannot suppose the whole medium to move with the molecules. In the equation I have put down, I want to guard against the supposition that it is a rigorously correct equation. In that equation, we supposed the molecules to be so evenly distributed that relatively to the dimensions of a thousandth part of a wave length if you like, it is practically a homogeneous solid, — in other words, an exceedingly fine grained solid, so finely grained that it is practically homogeneous for portions exceedingly small in linear dimensions in comparison with the wave length. But no degree of smallness will dispense with the to and fro motion of the elastic solid relatively to the imbedded molecules.

I want to invoke Lord Rayleigh, and if we can get him to take it up, we shall have a chance of learning something about it. I suppose the medium to move together with the imbedded molecules, as will be approximately the case if the effect of the molecules is such as to produce but a small difference in the circumstances from what they would be if there were no molecules imbedded at all. In other words, if the amount of this molecular action in the medium is such as to produce but a very small change in the velocity of light in proportion to the whole velocity, then I think we are quite clear in the assumption that the whole medium moves with the molecules. If in our formulas we put $C_1 = \infty$, $C_2 = 0$, etc., you will see that it is tantamount to adding the mass of M_1 to ρ the density of the medium which would not be the case unless the whole of the mass added increased the average density but little in proportion to the whole density. I think I am right in saying that if the medium becomes infinitely fine grained, and if the density is but little increased then the effect of putting in the molecules would be to add the mass per unit volume of the molecules to the density of the ether. I believe it is not so when the change of velocity is considerable in comparison with the velocity in the ether alone. And instead of our very nice, simple, mechanical arrangement that Prof. Rowland has illustrated for us here with springs between the rigid shell and the ether, it will give us an elastic action which will be playing to and fro among these molecules, and it will be a problem extremely difficult to solve, but since Lord Rayleigh has been induced to take it up, he will give us the answer. It is not absolutely a question for any bodies whatever or even spherical bodies. But it is the question, what kind of change in the equation we have put down will be introduced by taking into account that principal as to the motion of the luminiferous ether. I wanted to warn you against thinking for a moment that we can give fundamental value to the equations that I have put before you

We found $\mu^2 = 1 + \frac{C_1 \tau^2}{\rho}\left(-1 + \frac{q_1 \tau^2}{\tau^2 - \kappa_1^2} - \frac{q_2 \tau^2}{\kappa_2^2 - \tau^2} \cdots\right)$ When we have q_1 very nearly equal to 1, we can account for all we at present know of regular refraction, by values of τ greater than κ_1 and less than κ_2. If q_2 be excessively small, and τ not very much greater than κ_2, we may account for an absorption band as fine as you please. Suppose the question to be to account for refraction by vapor of sodium, not taking into account at present the double sodium line — that is to say, considering a substance like sodium, that gives only one line. Two terms I believe could be very reasonably arranged so as to give us, by the considerations we went through on yesterday, the irregular refractions that that medium would show. The period of this vapor would be κ_2. If q_2 is very small we shall have the absorption band appearing as a very sharp black line in the spectrum of the light coming through this vapor. This vapor put into a prism and experimented upon for the refractive power of the medium would give us something not distinguishable from ordinary refraction until you get near the period of the vapor, when there would be anomalous dispersion. But I say that if you take q_2 small enough, you may make the absorptional region and the region of anomalous dispersion as small as you please. I cannot doubt that this is the way the thing is done in nature; there is something in nature that corresponds precisely to that course of action.

I want you to think how small q_2 must be for the sodium line, thinking of only one sodium line. Sodium vapor shows no particular absorbing power until you have a period differing very little from the period of sodium vapor. How little you may judge by looking at the two sodium lines whose distance apart is about 1.001 and whose thickness is not more than $\frac{1}{50}$ of their distance apart. It is apparent from that that the dispersional region corresponds to a period say $T = \kappa_2 (1 \pm \frac{p}{50000})$ where p is a proper fraction. This third term must be insensible for values of T differing greatly from this. Therefore we must have $q_2 < \frac{1}{25000}$ according to these features.

Let there be a connection of particles so as to give another mode of vibration. It is not a hypothesis but a reality that sodium vapor has two independent period vibratcons whose periods differ by $\frac{1}{50000}$ of one another. We have then the means of making something which will modify the velocity of waves ~~through~~ jelly just as Sodium vapor modifies waves of light through the luminiferous ether. We have the means of making a mechanical model of the thing. I do not say it is the explanation of it.

Lecture XIV

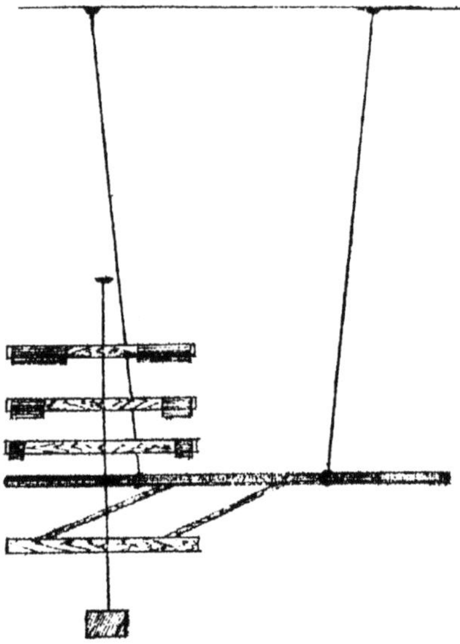

At this lecture, was seen immediately behind the model heretofore presented, two wires extending from the ceiling and sustaining a long heavy bar by means of closely fitting rings. By slipping these rings along the bar, the period of vibration about the bifilar suspension could be altered at will. Two pieces of wood served to transmit the motion of this vibrator to the lower bar P of the model.

This is another case from what I have been talking about. These rigid connections make the bar P go with a stated harmonic motion. I would like to have a heavy pendulum attached to the bar by a very light india rubber band. I want the vibrator to vibrate half a minute before you see any sensible motion in the model. This is another case likely, but it is quite equally interesting, and will do just as well. Let us look at this a little and see what it does. O, you can vary the period; that is very nice, that is beautiful. We are going to study these vibrations a little, just as illustrations. Prof. Rowland has kindly made this arrangement for us and I think we will all be interested in seeing it. We have this bar P, moved by this pendulum, this pendulum being so massive that its period is but little affected, I suppose, by being connected with P. It takes sometime before the initial vibrations in the model are got quit of and the thing settles into simple harmonic motion corresponding to the period of the pendulum. If we keep this pendulum going long enough through nearly a constant range the masses P, m_1, m_2, m_3, will settle into a definite simple harmonic motion, through the subsidence of any free vibrations which may have been superimposed upon them in the start. This now seems to be performing very nearly a simple harmonic motion. We will then superimpose another vibration on this by altering the period of the pendulum very slightly. That, you see, seems to have diminished very much the vibrations of the system. They are now increasing again. That will go on for a long time. I shall give this pendulum a slight impulse when I see it flagging, to keep its range constant. When it is in its middle position, I apply a working couple. We will give no more attention to it than just to keep it vibrating, while we look at these notes which I have prepared for you, so as to shorten our work upon the board.

Lecture Notes of October 13.

Homogeneous Elastic Solid of unrestricted character

(1) $e = \frac{d\xi}{dx}$; (2) $f = \frac{d\eta}{dy}$; (3) $g = \frac{d\zeta}{dz}$; (4) $a = \frac{d\eta}{dz} + \frac{d\zeta}{dy}$;

(5) $b = \frac{d\zeta}{dx} + \frac{d\xi}{dz}$; (6) $c = \frac{d\xi}{dy} + \frac{d\eta}{dx}$.

$W = \frac{1}{2}(e, f, g, a, b, c)^2 = \frac{1}{2}\{(e, f, g)^2 + 2(e, f, g)(a, b, c) + (a, b, c)^2\}$ (for brevity).

Problem I. given

$$\left.\begin{array}{r} 11, 12, 13, 14, 15, 16 \\ 22, 23, 24, 25, 26 \\ 33, 34, 35, 36 \\ 44, 45, 46 \\ 55, 56 \\ 66; \end{array}\right\} \text{"tansinomic" coefficients.}$$

required the bulk modulus (k). (NOTE, the "thlipsinomic" coefficients are more convenient for the case of incompressibility. They are more closely allied to practical moduluses.)

$$\left.\begin{array}{l} \text{In } (e, f, g)^2 \text{ put } e = e' + \tfrac{1}{3}\delta,\ f = f' + \tfrac{1}{3}\delta,\ g = g' + \tfrac{1}{3}\delta \\ \qquad \text{where } \delta = e + f + g \text{ and therefore } e' + f' + g' = 0 \end{array}\right\} \cdots (7)$$

$$\left.\begin{array}{l} \text{We find } (e, f, g)^2 = \tfrac{1}{9}[11 + 22 + 33 + 2(23 + 13 + 12)]\delta^2 \\ \qquad + \tfrac{2}{3}[(11 + 13 + 12)e' + (22 + 23 + 12)f' + (33 + 23 + 13)g']\delta \\ \qquad + (e', f', g')^2 \end{array}\right\} \cdots (8)$$

Hence $k = \frac{1}{9}[11 + 22 + 33 + 2(23 + 13 + 12)] \cdots (9)$

Problem II. To find W for the case of incompressibility we have $k = \infty$, $\delta = 0$, $k\delta^2 = 0$ ($k\delta$ may and generally is finite. Denote it by p. We don't need it now, but shall want it for equations of motion.)

Hence $W = \frac{1}{2}\{(e', f', g')^2 + 2(e', f', g')(a, b, c) + (a, b, c)^2\} \cdots (10)$

Problem III. (without restriction to incompressibility). To annul skewnesses relatively to OX, OY, OZ. This requires that, and is done when, $(e, f, g)(a, b, c) = 0$, and $(a, b, c)^2 = 44a^2 + 55b^2 + 66c^2$

Problem IV. (without restriction to incompressibility). To annul web-oidal aeolotropy, in the case of the annulled skewnesses.

In W take $e = \frac{1}{2}(r-s)$, $f = \frac{1}{2}(s-q)$, $g = \frac{1}{2}(q-r)$, we find

$$W = \tfrac{1}{2}\Big\{\tfrac{1}{4}\big\{(22+33-2.23)q^2 + (33+11-2.13)r^2 + (11+22-2.12)s^2 - 2\big[(11+23-13-12)rs + (22+13-12-23)sq + (33+12-23-13)qr\big]\big\} + 44a^2 + 55b^2 + 66c^2\Big\}$$

case $r = 0$, $s = 0$, shows that $\frac{1}{4}(22+33-2.23) = 44$
$s = 0$, $q = 0$, " " $\frac{1}{4}(33+11-2.13) = 55$ ....(11)
$q = 0$, $r = 0$, " " $\frac{1}{4}(11+22-2.12) = 66$

the necessary and sufficient conditions. They yield

$23 = \frac{1}{2}(22+33) - 2.44$; $13 = \frac{1}{2}(33+11) - 2.55$; $12 = \frac{1}{2}(11+22) - 2.66$. ....(12)

These, used in the coefficients of rs, sq, qr, give $\frac{1}{4}(11+23-13-12) = 55+66-44$, &c.

Thus finally

$$W = \tfrac{1}{2}\big\{44(q^2+a^2) + 55(r^2+b^2) + 66(s^2+c^2) - 2\big[(55+66-44)rs + (66+44-55)sq + (44+55-66)qr\big]\big\} \quad \text{....(13)}$$

N.B. This is for case of no dilatation. To find W without restriction, add to (13) the terms of (8) which involve δ.

I think it would be well to go through a rather full treatment of the problem of waves in an aeolotropic elastic solid. In preparation for it, we have to-day the dynamics of a homogeneous elastic solid of unrestricted character. I think perhaps it would have been better if instead of representing these taxinomic coefficients by 11, 12, etc., we had taken the notation of Thomson & Tait, (ee), (ef), etc. I would almost advise you to use (ee) instead of 11, etc.

There is a little note here to the effect that the thlipsinomic coefficients are sometimes more convenient than tasinomic coefficients. Tasinomic coefficients, in Rankines nomenclature, are coefficients of strain in formulas expressing stress. On the other hand, Thlipsinomic coefficients are coefficients of stress in formulas expressing strain. These coefficients can be got from one another by linear equations of course. We have for example a stress $P = (ee)e + (ef)f + (eg)g + (ea)a + (eb)b + (ec)c$ You have six equations like that for P, Q, R, S, T, U. The next will be $Q = (fe)e + (ff)f + (fg)g + (fa)a + (fb)b + (fc)c$, etc. $(ee), (ef)$ etc., are the tasinomic coefficients. Solve these equations for e, f, g, a, b, c. The thlipsinomic coefficients will be the coefficients $(PP), (PQ)$, etc., in the formulas $e = (PP)P + (PQ)Q + (PR)R + (PS)S + (PT)T + (PU)U$. etc. These are more convenient for working with incompressibility and are also more closely connected with the practical moduluses that we are familiar with. Young's modulus is the stress divided by the strain when the stress is a simple longitudinal force and the strain is an elongation connected with a contraction – a lengthening of the wire and lateral contraction of it. In the elementary experiment for Young's modulus, you apply a given weight and by observation find the elongation that that produces. The formula for Young's modulus is $e = (PP)P +$ terms which are zero, so that the reciprocal of a thlipsinomic coefficient $\frac{1}{(PP)}$ is Young's modulus.

Let us stop and look at this vibrating affair. It has been going a considerable time with the exciter going through a constant range and you see but small motion transmitted to the system. That is an illustration of the most general solution. Our handle P is in firm connection with the large pendulum ~~and~~ is forced to agree with it; and is to be viewed as the virtual exciter for a system of three particles. Let us bring these at rest. Now keep the pendulum going, and in the time when the viscosity will annul the system of vibrations, representing the difference between zero and the permanent state of vibration of these three particles, they will have acquired their permanent vibration. If there were no loss of energy whatever, the result would be that this jangled state would last forever, consisting of a simple harmonic motion in the vibrator and a compound of the three fundamental modes of these three particles viewed as a vibrating system with this bar P held fixed. Set this system with the lowest bar P held fixed to vibrating in any way whatever, and its motion will be a compound of those three fundamental modes. Besides that, set this exciter going and the state of the case is this: we may have this exciter and the whole set in simple harmonic motion of the same period, or superimposed upon that, any composition of the three sets of vibrations that the system might have with the exciter fixed. We cannot improve on the mathematical treatment by observation; and really a thing of this kind is more as a help or corrective to brain sluggishness than as a means of observation or discovery. In point of fact, we can discover a great deal better by algebra. But brains are very poor after all, and this model is of some slight use in the way of making plain the meaning of the mathematics we have been working out.

The system seems to have come once more into its permanent state. Let us stop this vibrator and see how long the system will hold its vibration. The reaction of the exciter is very slight, it is very nearly the same as if that bar were absolutely fixed. But the motion communicated to it since it is not absolutely fixed will correspond to a considerable loss of energy. A very slight motion of that bar with its great length and weight has considerable energy compared even with the energy of our particle of greatest mass, so that this system will come to rest far sooner than if this bar were absolutely fixed. These particles are at present illustrating phosphorescence. You see they have gone on vibrating for a whole minute, and the lowest of these three bars must have performed a couple of dozen vibrations at least. A phosphorescence of a hundred seconds duration is quite analagous to the giving back of vibrations by that system (for two or three dozen vibrations only instead of two or three dozen vibrations we have 40,000 million million vibrations during the hundred seconds. Now, we cannot get 1000 vibrations out of this system, because of the loss of energy in the wire, resulting from the generation of heat in it, (which in our mind's eye we can see very clearly is connected with this system and is running away with its energy). That generation of heat by viscosity, is simply the conversion of energy from one state of motion into another. In our molecular dynamics, we have no underground way of getting quit of energy or carrying it off. We must know exactly what is done with it when the vibrations end after a thousand million million. We must suppose the elasticity of our matter and molecules and so on to be perfect; and we cannot in any part of our molecular dynamics admit unaccounted for loss of energy; that is to say we cannot admit viscous terms unless as an integral result of vibrations connected with a part of the system that is not convenient for us to look at.

In three minutes our system has come very

nearly to rest. We infer, therefore that in three or five minutes from the commencement of a vibration we shall have nearly the permanent state of things.

Now we vary the period of the exciter making it as unlike any fundamental period as possible. We will keep this going in an approximately constant range for a while and look at the vibrations that produces in the system.

I have explained how the thlipsinomic coefficients are more closely allied to practical moduluses. I may say, however, that in point of fact one of our taxinomic coefficients is the pure modulus of rigidity in an isotropic body; but it may be regarded as the reciprocal of the corresponding thlipsinomic coefficient. Take the quadratic function in a, b, c which I express shortly $(a, b, c)^2 = \frac{1}{2}\{44a^2 + 55b^2 + 66c^2 + 2(56bc + 46ac + 45ab)\}$. In the case of an isotropic solid $44 = 55 = 66 = n$, the rigidity; and the taxinomic equation is $S = na$ ∴ the thlipsinomic equation is $a = \frac{1}{n}S$, and the reciprocal of the rigidity is a thlipsinomic coefficient. The taxinomic and thlipsinomic coefficients for an isotropic body are not reciprocals of each other however in the case of longitudinal strains. You may readily see that the two are not reciprocals in any case in which there is more than one term in the linear function by which the stress or the strain is expressed.

Now you see very markedly the difference between the vibrations of our system after it has been going for several minutes with the exciter on a somewhat shorter period of vibration than that with which we commenced. Here is another still shorter. In the course of two or three minutes the superimposed vibrations will die out. See now the tremendous difference of this case in which the period of the exciter is approximately equal to one of the fundamental periods of the system, or the periods for the case in which the lowest bar is held absolutely fixed. The angle through which that bar turns corresponds to ξ in our formula.

Returning to our tasinomic expression, required the bulk modulus. [Problem I.] Taking for the moment average pressure per unit of area all around — for instance on the three pairs of faces of a cube — as the stress, the bulk modulus is $\frac{\frac{1}{3}(P+Q+R)}{e+f+g}$. We may obtain the solution in this way: Let the actual elongations be represented in terms of elongations e', f', g' which produce no change of bulk, and δ, as in the notes before you. The work required to produce the state of things represented by $e = e' + \frac{1}{3}\delta$, etc., will be the term in δ^2 in $\frac{1}{2}(e,, f, g)^2$. Let k be the bulk modulus, and consider the work done in distortion. The working pressure varies from nothing to p, the final pressure, which, according to our definition of the bulk modulus $= k\,\delta$. The average working pressure therefore $= \frac{1}{2}k\delta$, and the work done $= \frac{1}{2}k\,\delta^2$. Therefore k is equal to the coefficient of δ^2 in $(e, f, g)^2$.

For the particular case of an isotropic body we have $11 = 22 = 33 = \mathfrak{A}$, $12 = 23 = 31 = \mathfrak{B}$ $\therefore$ $k = \frac{1}{3}(\mathfrak{A} + 2\mathfrak{B})$. That then, coming down to the particular case of an isotropic body is the relation between the tasinomic direct modulus, and the tasinomic lateral modulus. To interpret these, let everything $= 0$ except f. Therefore $P = \mathfrak{B}f$ which means that $\mathfrak{B}$ is the force per unit of elongation in the direction perpendicular to the force. Again, $Q = \mathfrak{A}f$, which means that $\mathfrak{A}$ is the force per unit of elongation in the direction of the force. These are moduluses that we are not so familiar with in practice.

So much then for our first problem. Next to find W for the case of incompressibility. This is a somewhat difficult conception to deal with, since every one of our coefficients are infinite. For the case of incompressibility, we put $k = \infty$ $\delta = 0$ $k\delta^2 = 0$; $k\delta$ generally remains of finite magnitude and will take the place of pressure. In the case of an isotropic body, $k\delta$ is the average pressure. Putting this compressibility modulus $= \infty$, into the form of an equation, we have $11 + 22 + 33 + 2(23 + 13 + 12) = \infty$. Except in some special and exceptional case, each one of these 6 quantities 11, 12, etc. will be infinite. But the ratios between them that are effective in the expression for the energy in the case of a pure distortion of the solid in question, are finite. It is upon this account that the thlipsinomic coefficients are more convenient in the case of incompressibility. We can scarcely treat an algebraic equation of 21 quantities, each infinite, with the finite ratios between them not explicitly stated; so that we are left in a doubtful position.

Now let us look at problem IV. Without restriction as to incompressibility – with none of these infinities – to annul web-oidal aeolotropy – "Weblike" I should say. I have not a Greek Dictionary with me, and have not the grand command of Classical knowledge which Rankine had. Every one of Rankine's words are well chosen and it is a most instructive lesson on the theory of elastic solids just to read them over. I want something for "web". Can any one tell us the Greek for web? Well, weblike then. That is the kind of aeolotropy we have in a piece of woven cloth. I introduce a temporary notation in the quadratic expression for the work, $e = \frac{1}{2}(r-s)$, etc. This assumes e, f, g to be such as to give no change of bulk. I am not assuming that the solid is incompressible, but I am assuming the case of distortion without change of bulk. The most convenient way of expressing that is to take three quantities q, r, s, and put e, f, g, equal to their differences so that we have $e+f+g=0$. You might express e, f, g in terms of two quantities by means of this relation but that is unsymmetrical. The symmetrical system is a great brain saving system in all cases in which it is useful.

I would be much obliged if mathematicians would verify this work. To understand it take the particular case $r=0$, $s=0$, so that there remains only q. What is the meaning of q in this case? We have $e=0$, $f=-\frac{1}{2}q$, $g=\frac{1}{2}q$. It was the "half business" coming in here that was my reason and justification for the notation in my paper on elasticity that I referred to, which I am not insisting upon at present. But I will give you a reference to-day to Thomson & Tait, Art. 681, which will show you the importance of the question that I answered in a very special way which unfortunately becomes too artificial in this case. The marginal statement is, "Discrepant reckonings of shear and shearing stress, from the simple longitudinal strains or stresses respectively involved". The question is to pass from positive and negative normal pressure perpendicular to two diagonal planes to the reckoning of simple stress. The reckoning of simple stress is simply the amount of tangential traction in either set of planes. On the other hand, the numerical measure of the shear or simple distortion comes out double the amount of elongation or contraction in the diagonal analogue. To make them both the same, I put in a $\sqrt{2}$. For this particular application it is not worth while to do that; but in the system set forth in that paper on Elasticity we have a convenient symmetrical method of reckoning all stresses and strains so that the resultant of two orthogonal shears shall be the square root of the sum of their squares.

Take then the particular case $r=0$ $s=0$. What is the amount of shear corresponding to f, elongation in one direction, and g, contraction in the other? Answer, $2f$, $2g$; that is to say, g measures the shear corresponding to $\frac{1}{2}g$ elongation in one direction and $\frac{1}{2}g$ contraction in the other. The two directions are OY, OZ. We have an extension in the direction OY and a shortening in the direction OZ and the question is, "What is the simple shear corresponding to that." ~~and the~~ and the answer is "it is numerically equal to twice the elongation, or to g." Thus g is the strain in the plane perpendicular to OX; but $\underline{a}$ is the strain in the plane perpendicular to OX ∴ the coefficient of g^2 in the equation of energy for this particular case must be equal to the coefficient of $\underline{a}^2$ or $\frac{1}{4}(22+33-2.23)=44$, which is the formula stated. That condition is to express that there is such a deviation from aeolotropy as would be produced if we were to annul the differences of rigidity relatively to a shear produced by pulling out one diagonal and shortening the other compared with the shear of sliding one face past the other.

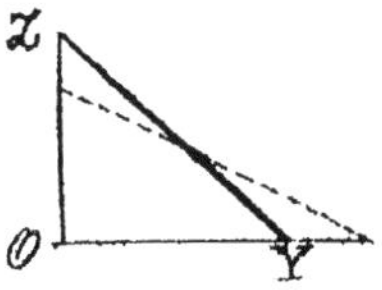

Suppose now you want to get quit of the sidelong coefficients 12, 13, 23. This equation, $\frac{1}{4}(22+33-2.23)=44$, you see expresses 23 in terms of 22, 33, 44. These equations used in the coefficients of $r.s$, sq, qr, give $\frac{1}{4}(11+23-13-12)=55+66-44$, etc.; and there remains finally, for the energy, the expression ~~marked~~

(13). In that expression for the energy, we have everything expressed in terms of 44, 55, 66, the three principal rigidities, and altogether independent of the moduluses that express the effects of direct pressures. We have here the most general kind of distortion; and we have the work of that distortion expressed in terms of the three rigidities; and we are ready, therefore to go on and investigate distortional waves without further question as to whether the elastic solid is compressible or not. That question will only come up when we get to the reflection or refraction of light at the bounding surface of two mediums; or when we put in our molecules, or introduce equations which would produce condensation or rarefaction in the medium. But for the present we do not want to consider, whether it is compressible or not; and that, in point of fact, is Green's position.

I had almost hoped that I would see some way of explaining double refraction by this system of molecules, but it seems more and more difficult. I will take you into conference to-morrow, if you like, and show you the difficulties that weigh so much upon me. I am not altogether disheartened by this, because of the fact that such grand and complicated and highly interesting subjects as I have named so often, absorption, dispersion and anomalous refraction, are all not merely explained by their means but are the inevitable results of this idea of attached molecules.

There is one thing I want to say before we separate and that is, when I was speaking last of the subject, I saw what seemed to me to be a difficulty but on further consideration, I find that there is no difficulty at all. Not very many hours after I told you it was a difficulty I saw that I was wrong in making it appear to be a difficulty at all. I do not want to paint the thing any blacker than it really is and I want to tell you that that question I put as to the ether keeping straight with the molecules is easily answered when there is a large number. Our assumption was a large number of spherical cavities, lined with rigid spherical shells and masses inside joined by springs or what not, with the distance from cavity to cavity small in comparison with the wave length. It then happens that the motion of the medium relatively to the rigid shells will be exceedingly small and a portion of the medium that will contain a large number of these shells will all move together. If the distance from molecule to molecule is very small in comparison with the wave length then you may look upon the thing as if the structure were infinitely fine, and you may take it that the ether moves quite straight with them all, and not in and out among them as I said. It is evident that when the wave length of the medium is moderate in comparison with the distance between the particles that it can move out and in among them. But if the stiffness of the medium is such as to make the wave length large in comparison with the distance from molecule to molecule the stiffness is sufficient to keep them all together, and you may regard these rigid shells as bonds of attachment by which the molecule is pulled this way and that way, and that the millions and millions of these present the same effect as if the medium were made denser, so that we may suppose our reactionary forces, of which $c_1(\xi - x_1)$ is a sample to be absolutely the same in their effect upon the medium as if they were uniformly distributed through it.

That takes away one part of the dissatisfaction to the thing. The only difficulty that I see just now is that of explaining double refraction. The subject grows upon us terribly, and so does the time, I think. If it is not too much for you I must have one of our double lectures to-morrow.

Lecture XV.

We shall have in a short time a state of things in this model not very different from simple harmonic motion, if we get up the motion very gradually. We have now an exciting vibration of shorter period than the shortest of the natural periods. We must keep the vibrator going through a uniform range. We are not to augment it; and it will be a good thing to place something here to mark its range. Keep it going long enough and we shall see a state of vibration in which each bar will be going in the opposite direction to its neighbor. If we keep it going long enough we certainly will have the simple harmonic motion; and if this period is smaller than the smallest of the three periods, we shall, as we know, have these bars going in opposite directions. There is a longer period vibration of the largest mass superimposed on the simple harmonic motion we are writing for. I will try and help to that condition of affairs by resisting that vibration of the top particle. In fact, that particle will have exceedingly little motion in the proper state of things, (that is to say, when the motion is simply harmonic throughout) and it will be moving so far as it has motion at all in an opposite direction to the particle immediately below it. It is nearly quit of that superimposed motion now. We cannot give a great deal of time to this, but I think we may find it a little interesting as illustrating dynamical principles. Prof Mendenhall is here acting the part of an escapement in keeping the vibrator to its constant range. We cannot get quit of the slow vibration of the particle. A touch upon it in the right place may do it. A very slight touch is more than enough. I have set it the wrong way.

Prof. Morley has been so kind as to work out a large part of the solution of this problem for the seven particles that I gave you, so that we shall be able to see the distribution of energy among the masses in the different modes of vibration, and so get a very instructive lesson, as I believe, in respect to fluorescence, phosphorescence, and the radiation from a body which has become heated by the transmission of radiant heat through it; Now we have got quit of that vibration and you see no sensible motion of the upper particle at all; these two are going in opposite directions, the lower one going in an opposite direction to the exciter. Therefore this is a shorter vibration than the shortest natural period. Now I set it to agree with the shortest of the periods, the first critical position. If we get time in the second lecture to-day, I am going to work upon this a little to try to get a definite example illustrating a particle of sodium.

Before we enter upon any hard mathematics, let us look at this a little, and help ourselves to think of the thing. What I am doing now is very gradually getting up the oscillation. I am doing to that system exactly what is done to the sodium molecule, for example, when sodium light is transmitted through the vapor. We may feel quite certain, however, that the energy of vibration of the sodium molecule goes on increasing during the passage through the medium of at least two-hundred thousand waves, instead of two dozen at the most perhaps that I am taking to get up this oscillation. But just note the enormous vibration we have here, and contrast it with the state of things that we had just before. The upper particle is in motion now and is performing a vibration in the same period and phase as the lower particle, only through comparatively a very small range. The second particle, I am afraid, will overstrain the wire. By hanging up a watch, bifilarly, so that the period of bifilar suspension approximately agrees with the balance wheel, you get likewise a state of wild vibration. But if you perform such experiments with a watch, you are apt to damage it. This is a most magnificent contrast to the previous state of things when the period of the exciter was very far from agreeing with any of the fundamental periods.

We will now go to the treatment of the elastic solid. You will see a note in the paper of yesterday to which I have referred, stating that the Ellipsoidal method is more convenient for dealing with incompressibility, and in point of fact it is so. I feel certain that if the k given by formula (9) is $=\infty$, that the body must be incompressible, but that is the sum of six quantities each of which is generally—

I believe essentially — positive for a true elastic solid. May some of these be finite, or is each one infinitely great? In all ordinary cases each one of the six quantities is infinitely great, and we are left in an unsatisfactory state as to coefficients. It will be necessary to go through a good piece of analytical work to make this clear and satisfactory. This is well worth doing, but we have not time to do it. Any of you who may wish to go into it, may proceed thus: Express the 21 coefficients in terms of 19 coefficients and k, which you can do by algebraical processes. Suppose k very great and see how things get on; then suppose k infinitely great and I think you will get some reasonable expression for incompressibility in terms of taxinomic coefficients.

I explained to you yesterday Rankines nomenclature of thlipsinomic and taxinomic coefficients. In a certain sense, these may be all called moduluses of elasticity. I have defined a modulus as a stress divided by a strain, following the analogy of Young's modulus. If we adhere to that then the taxinomic coefficients are moduluses, and the thlipsinomic coefficients are reciprocals of moduluses. The relations between the taxonomic and thlipsinomic coefficients are well worked out by Rankine, but you can all do it for yourselves by going into the algebra concerned. There is not time for us to go into these matters in much detail. What we want is the essence of the dynamics. As far as symbols help us to that we shall use symbols; and when symbols do not help us to that, we shall let them alone. We will now look at our paper:

Lecture Notes, Oct. 14

Thlipsinomic discussion of compressibility and incompressibility

$$\left.\begin{aligned} e &= (PP)P + (PQ)Q + (PR)R + (PS)S + (PT)T + (PU)U \\ f &= (QP)P + (QQ)Q + (QR)R + (QS)S + (QT)T + (QU)U \\ g &= (RP)P + (RQ)Q + (RR)R + (RS)S + (RT)T + (RU)U \end{aligned}\right\} \cdots (15)$$

Hence

$$(e+f+g), \text{ or } \delta = [(PP)+(QP)+(RP)]P + \dots + [(PT)+(QT)+(RT)]T + \dots \quad (16)$$

Thus we see that $[(PP)+(QR)+(RP)], \dots [(PT)+(QT)+(RT)], \dots$ of this formula are six compressibilities.

And for incompressibility each must be $=0$, giving six equations,

$$\left.\begin{aligned} [(PP)+(QP)+(RP)] &= 0 \\ \text{-------------}& \\ [(PT)+(QT)+(RT)] &= 0 \end{aligned}\right\} \quad \dots\dots\dots \quad (17)$$

Case of annulled skewnesses (Prob. III, of Oct 13). The necessary and sufficient conditions are

$$\left.\begin{aligned} (PS) &= 0, & (QS) &= 0, & (RS) &= 0 \\ (PT) &= 0, & (QT) &= 0, & (RT) &= 0 \\ (PU) &= 0, & (QU) &= 0, & (RU) &= 0 \\ (TU) &= 0, & (US) &= 0, & (ST) &= 0 \end{aligned}\right\} \quad \dots\dots\dots \quad (18)$$

(Twelve annulments leaving nine coefficients)

In this case three of the compressibilities are annulled. The others are:

$$(PP)+(QP)+(RP), \ (QQ)+(RQ)+(PQ), \ (RR)+(PR)+(QR) \quad \dots \quad (19)$$

It is startling to think of six equations to express incompressibility, I have not really noticed it before, but it is quite right, and you see the reason for it in this way: Consider an absolutely aeolotropic solid without any limitation whatever. Take this model of an elastic solid, if you like, that I showed you the other day, with its 18 coefficients. We will apply opposite shears to it. I shall apply a couple in this direction, and Mr. Forbes will balance that with a couple in that direction. Every one of you can understand the sort of thing that that does to the box. Suppose the axis of X is vertical. What we are doing is to shear this in the plane Y Z by shears parallel to the axes. If the body be absolutely aeolotropic, doing this will alter its bulk; and again, to alter its bulk will produce that shearing effect.

Rankine did a great deal to cure the mathematical disease of aphasia from which we suffered so long; Faraday did most. The old mathematicians used neither diagrams to help people to understand their work, nor words to express their ideas. It was formulas and formulas alone. Faraday was a great reformer in that respect with his language of "lines of force," etc. Rankine was splendid in his vigor, and the grandeur of his Greek derivatives. Perhaps he over did it, but I do not like to call it an error. We cannot use all his words, but we learn from them in reading his papers. Instead of his platytatic and platythliptic coefficients, I use the much less grand and more colloquial expressions, sidelong normal and sidelong tangential coefficients. I do not know that Rankine has a word for the interaction between shears and shearing forces parallel ~~to~~ to the faces, and direct strains. A direct strain in this case is an elongation parallel to any of the three axes. I assume you know what that means. These cross connections between shears and distorting stresses on the one hand and normal forces and a simple dilatation on the other, I can talk of as sidelong.

Look now at (15). What does (PS) mean? It expresses a relation between a distorting stress S, such as that which Mr. Forbes and I apply, and a shrinkage

P. Annul everything in (15) except S, and the result is $e = (PS)S$, $f = (QS)S$, $g = (RS)S$, so that (PS) is the dilatation we are causing in the direction OX. $(PS)+(QS)+(RS) = 0$ means that there is no dilatation from what we are doing. It is clear, therefore, by this, that we have six equations to express that there is no dilatation under any kind of stress. You see also, how readily one is led to the treatment of incompressibility by thlipsinomic coefficients, while on the other hand, it is very troublesome in terms of the tasinomic coefficients.

We next take up the case of annulled skewnesses — using a gross word as you see. I forget what Rankine's word for that would be. Skewnesses is a common word, but it is sanctioned by great mathematician so that we need not be ashamed of it. The annulment of skewnesses is set forth in problem III (Oct. 13) in short language, $(e, f, g)(a, b, c) = 0$, which means, of course, that the cross coefficients $(e\,a) = 0$, $(e\,b) = 0$, $(e\,c) = 0$, $(f\,a) = 0$, etc. That means 9 equations then, written short. Those 9 equations are obviously essential for annulment of skewnesses. Three more equations are necessary, viz: the sidelong coefficients $(b\,c) = 0$, $(c\,a) = 0$, $(a\,b) = 0$, so that the quadratic $(a\,b\,c)^2$ reduces to a sum of squares.

To explain the tasinomic conditions the question put is, what stress is required to produce a stated strain. Let, for example, the stated strain be a shear in the plane YZ denoted by $\underline{a}$. If the body be aeolotropic, a stress compounded of P, Q, R, S, T, U will be required to produce it, none of the coefficients vanishing. But if the body be free from skewnesses, then it is clear that a shear of this kind requires no stress to produce it except the one corresponding to this shear. That is to say, shear $\underline{a}$ is produced by stress S, shear $\underline{b}$ is produced by stress T, shear $\underline{c}$ is produced by stress U. Therefore we have 12 equations in order to annul skewnesses, bringing us down from 21 coefficients to 9. Why do we not, in avoiding skewnesses, annul the sidelong coefficients (ef), (fg), (eg)? We do not, because obviously, without any skewnesses, a strain in this direction, requires a stress in directions at right angles to it to prevent the body from swelling or contracting in those directions. Therefore (ef), (fg), (eg) belong clearly to the non-skew system, so that we have essentially 9 coefficients in that system.

To annul web-like aeolotropy requires the three equations (11). What may be taken as the most convenient yield of the problem are equations (12), because they allow us to get quit of the sidelong coefficients (ef), (fg), (eg) [= 12, 23, 13], leaving the direct coefficients 11, 22, 33, and the three principal rigidities 44, 55, 66. These equations (12) are of some importance. They are three out of Greens five equations by which he expresses that of the three possible waves having wave front in one plane, two consist of vibrations in that plane, and one of vibrations perpendicular to it. His other two are 11 = 22 = 33. That shows you exactly the relation between Green's equations and the results that we have arrived at by the practical and static consideration of an elastic solid. I suppose most of you have Green's collected papers. I ask you the question because we shall use it a little in what follows. You will find these 5 equations at the foot of page 302.

We are about ready for the wave surface, but this is not an elementary class and we will not go into the geometry of the wave surface but will think of the results. As I said in the first lecture one of the difficulties is quite refractory indeed. In the wave theory of light the velocity of the wave ought to depend on the plane of distortion. If you compare the results of the wave surface worked out for an incompressible aeolotropic elastic solid (we shall look at that a little more presently) you will see that it agrees exactly with Fresnel's wave surface if instead of the direction of the line of vibration of the particles in Fresnel's construction we have the normal to the plane of distortion.

I see no way of getting over the difficulty that the return forces in an elastic solid — the forces on which the vibration depends — are dependent on the strain experienced by the solid and on that alone. There has always seemed to me something indigestible in the way Green gets over it. I see that Stokes quotes in his Report on Double Refraction, page 265 (British Association 1862). "In his paper on Reflexion Green had adopted the supposition of Fresnel that the vibrations are perpendicular to the plane of polarization. He was naturally led to examine whether the laws of double refraction could be explained on this hypothesis. When the medium in its undisturbed state is exposed to pressures differing in different directions, six additional constants are introduced into the function ϕ, or three in the case of the existence of planes of symmetry to which the medium is referred. For waves perpendicular to the principal axes, the directions of vibration and squared velocities of propagation are as follows:—

Wave normal		x	y	z
Direction of vibration	x	$G+A$	$N+B$	$M+C$
	y	$N+A$	$H+B$	$L+C$
	z	$M+A$	$L+B$	$I+C$

"Green assumes, in accordance with Fresnel's theory, and with observation if the vibrations in polarized light are supposed perpendicular to the plane of polarization, that for waves perpendicular to any two of the principle axes, and propagated by vibrations in the direction of the third axis, the velocity of propagation is the same."

We will try and keep this last sentence in our heads and study it. I have had an exceedingly exciting time since I saw you yesterday. I could not swallow this. It seemed to me to be absolutely wrong.* I feel this to be a very serious statement to make when Stokes quotes it and says that Cauchy does the same thing.

Let us see what this statement means before considering whether it may be verified as Green supposes, by the introduction of "extraneous pressures." We are to have waves (for example N and W) perpendicular to any two of the principal axes, each propagated by vibrations in the direction of the third axis (up and down). Take first the wave that is propagated South as I hold the box.

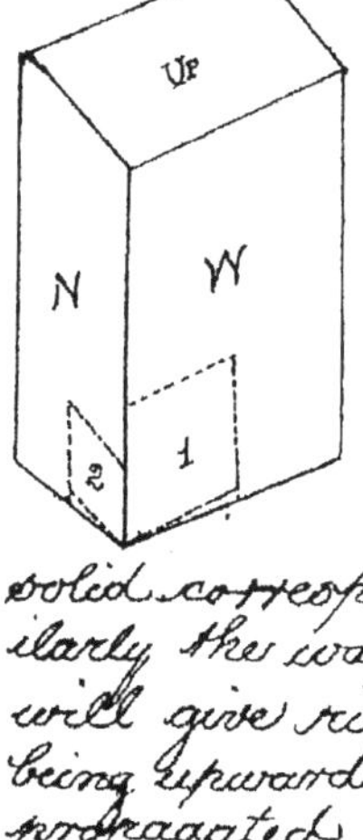

There is the plane of the wave (N). The vibration up and down will consist of a distortion in this West plane (W). An upward vibration will give a shear like that (1) in which a rectangular figure becomes a rhombic figure. That represents the strain in the solid corresponding to this first state of motion. Similarly the wave propagated in this eastward direction will give rise to a shear of this kind (2), the vibration being upward. The assertion is that one set of waves is propagated at the same speed as the other. That is to say,

*[I am in receipt of a letter from Sir Wm Thomson stating that he has thought of "extraneous forces" which can give rise to return forces dependent on rotational displacements, so that Green is here correct. The letter will be incorporated in the conclusion of this discussion of Green's second theory in Lecture of Oct. 15. H.]

the waves which have their shear in this west plane have the same velocity as the waves which have their shear in this north plane. The essence of our elastic solid is three different rigidities one for shearing in this plane W, one for shearing in this plane N, and one for shearing in the other principal plane. The assumption is then that the velocities of propagation are the same in planes having different shears, i. e., do not depend on the shearing strain.

The introduction by Green (in order to accomplish this) of what he calls "extraneous force" which gives him three other coefficients has always seemed to me of doubtful ingenuity. These coefficients A, B, C, occur in the little table given above and L, M, N, are the three principal rigidities. The table gives the squared velocities of propagation and waves of different wave normals and directions of vibration along the axes. The principal diagonal refers only to condensational waves, or waves in which the direction of vibration coincides with the wave normal. Taking vibrations in the direction x, the assertion is, $N + B = M + C$, which with the two corresponding equations for vibrations in the directions y, z, lead to $A - L = B - M = C - N$.

A, B, C are the effects of extraneous pressure. So far as I can see, they must be null. Begin with a body quite isotropic, so that we may not have our minds confused with the complicated question of aeolotropy — an elastic jelly, say, in a rectangular box. Let the box be altered in shape, still retaining its rectangular form. Will there be any difference of elasticity produced? Certainly not. The superposition of displacement will go on just as though the displacement and the external forces forming a system in equilibrium did not exist. Write down the equations if you like, expressing a stress in any portion of the solid. Superimpose a ~~[illegible]~~

previous strain and you simply add to the formula the expression for the previous strain. The expression for stress in terms of strain is not modified by the fact that you superimpose a stress and strain upon a stress and strain previously existing. I say, therefore it is a mistake to introduce the coefficients A, B, C if they correspond to nothing in nature." Make these annulments of A, B, C, and there is a table and a very convenient one you may find it for the squares of the velocities in the directions stated for wave normals and vibrations. These quantities in our notation are G = (ee), H = (ff), I = (gg), L = (aa), M = (bb), N = (cc).

I have said to myself, is it possible, after all, that this refractory difficulty can really be got over by that supposition of an extraneous force. I would not be lamenting that we could not explain double refraction if that were so; for this has long been a form of the elastic solid theory by Green in which he gets reconciliation to Fresnel's construction.

I will read two or three passages from Green. We may go back to where this is first mentioned in Green's paper On the Reflexion and Refraction of Light. On page 248, he says, "Let us conceive a mass composed of an immense number of molecules acting on each other by any kind of molecular forces, but which are sensible only at insensible distances, and let moreover the whole system be quite free from all extraneous action of every kind." That is what Green supposes first, and again he says (page 250), "The formula just found is true for any number of media comprised in this volume, provided the whole system be perfectly free from all extraneous forces, and subject only to its own molecular actions."

All this first paper is absolutely right, except the logic of those two passages that I have quoted, viz, "provided the whole system be perfectly free from all extraneous forces. If I am right in saying that the effects of his extraneous forces are null; that is logically wrong. If it is logically right the error is mine. He uses that logic in his paper On the Propagation of Light in Crystallized Media read May, 1839, and in the very last paper in the book On the vibration of Pendulums in fluid Media, read at a considerable earlier date, Dec. 1833. The way he introduces it (and I have always turned from it when I saw it) is, (p. 298) "If there were no extraneous pressures, the supposition that the primitive state was one of equilibrium would require $\phi_1 = 0$, as was observed in the former paper; but this is not the case if we introduce the consideration of extraneous pressures."

Green meant a proposition of which this is a sample: take an elastic jelly; elongate it in one direction and shorten it in directions at right angles to that, and that will produce aeolotropy, introducing difference of propagation of waves in different directions in the manner his formula would show, with the A, B, C coming in along with the L, M, N. Here is how you might introduce aeolotropy into a jelly: viz., by compressing it beyond the limits of its elasticity. That is quite another affair. Even then, you just make it a crystalline solid, and you will come back upon a case in which the velocity of propagation will depend on the direction of the strain in this previously isotropic solid which has been rendered aeolotropic by stress. This mode of aeolotropy, fulfils other than the conditions Green wants.

What the result of the introduction of extraneous force may be is of very great importance. If it be what it seems to me it is, it cuts away the last ground for explanation of the propagation of waves in a strained or unstrained elastic solid so as to fulfil Fresnel's law that the velocity of propagation depends on the direction of vibration rather than upon the plane of distortion.

As for the future of this work, I almost think it will pay us better not to trouble ourselves much more about wave surfaces. It is very pretty geometry; and if we had another week or fortnight we might do more upon it. I may give you another leaf like this, putting down the wave surface on the elastic solid theory. But what we want to do is to think of the wave surfaces that we may get by other conceivable suppositions, suppositions that make the velocity of propagation depend on something else than the distortion of an elastic medium; and to think of whether by any of these methods, we can get a wave surface agreeing with Fresnel's absolutely, or as nearly as the limits of accuracy of observation on which the belief in Fresnel's wave surface is founded, require.

Before leaving this, I want you to notice that our equations of yesterday bring us virtually down to the assertion that when the wave is one of distortion alone and when the solid is symmetrically related to the axes, i.e., when we have got rid of skewness and web-like asymmetry the problem is reduced to dependence on the three rigidities, so that all we want to know is 44, 55, 66. The 11, 22, 33, disappear either in the compressibility affair or in the condensational wave which may be propagated independently of the distortional wave in a true elastic solid. We do not care to get quit of the condensational wave so far as the the theory of waves in crystals is concerned. It is only when we come to the subject of reflection and refraction that we require the conditions of incompressibility.

Lecture XVI.

I want to call your attention to this, that Green's formula for the energy on page 299 expresses the energy virtually as a function of strain components and rotation components. It is not explicitly put in terms of rotation components; it is put in terms of of true strain components and certain differential coefficients which are neither pure strains nor pure rotations. I hope to get something written out on that by to-morrow, to put in your hands, to show precisely what Green's formula means, and you will see that it express energy in terms of rotation, as well as strain. If you think of the thing physically, you will see, I think, that it is quite impossible that a portion of the solid has been turned around by the extraneous forces. There is no relation to any non-rotating body by which we can possibly get terms in the potential energy depending on rotations. If there are terms in the potential energy depending on rotations there would have to be terms in the expression for the forces required to hold the body displaced depending on the angle through which a portion is turned and that is obviously not the case.

Green does not discuss his energy formula as we are accustomed to do. He had not risen to the ideas of potential energy and the systematic interpretation of the coefficients that are now so familiar to us. He is one of those who led the way, but who died before going so far on it as has been done by his successors.

Now I want to think a little more about the possibility of explaining the phenomena of light by our system of detached molecules. As we have been touching so near upon double refraction, I shall continue upon it, and show you my difficulty as I promised. If time permits, in the few days that remain, we shall put down a little more definitely, perhaps, the wave surface, and so on, that we are led to by such aeolotropy as we can get. I want, somehow or other, to extort an aeolotropy which shall be available for double refraction, out of our supposition of molecules imbedded in an isotropic medium.

Take for our detached molecule the very simplest case of one particle, m_1. This is equivalent to making the remote attachment of spring c_2 a fixed point, or to making $m_2 = \infty$ in our equations. We thus find directly $\frac{x_1}{\xi} = \frac{c_1}{c_1 + c_2 - \frac{m_1}{\tau^2}}$, which substituted in the formula for the square of the refrangibility gives $\mu^2 = 1 + \frac{c_1}{\rho}\left(\frac{x_1}{\xi} - 1\right)\tau^2 = 1 + \frac{c_1 \tau^2}{\rho}\left(\frac{\frac{m_1}{\tau^2} - c_2}{c_1 + c_2 - \frac{m_1}{\tau^2}}\right)$. The period of the molecule is given by $\kappa_1^2 = \frac{m_1}{c_1 + c_2}$. If τ^2 lies between $\frac{m_1}{c_1 + c_2}$ and $\frac{m_1}{c_2}$, all that I have said in favor of the more general expression, with reference to its availability for representing in a reasonable manner the facts of the ordinary refraction apply as well to this; but in the former we can help ourselves, if necessary in any case to explain the facts of refraction, by a critical period considerably greater than the longest period with which we have to deal. It is probable that if we go into the thing very fully, examining such results as Langley's with rock salt, etc, we shall have need of something of that kind. There is not the same wealth of coefficients in this to explain the observed vibrations of refraction that we have in the general solution; but I do not know that we can get much out of the general solution that we cannot get out of this, so far as ordinary refraction is concerned.

Our supposition is that a smaller velocity of propagation than in the luminiferous ether is due to mole-cules being attached by a something to the ether. If it is explained by imbedded molecules, difference of velocity for waves in different directions (or in other words double refraction) must be explained in the same way. Let us try to do so. First μ^2-1 must be something that is nearly constant for variations of T. The greatest dispersion is from 1.5 to 1.6; 4% of difference in velocity from extreme red to extreme violet is very high dispersion; for ordinary refraction the difference in most cases is not more than a few per cent. On the other hand, there is a little difference between the double refractions (in iceland spar we have, for the ordinary and extraordinary ray 1.4 and 1.6, which shows at once a difference of $\frac{1}{7}$ between the two refractive indices) – but double refraction is not a phenomenon of prismatic colors, and the difference between the two refractive indices for the extreme cases in iceland spar, although it does differ for the different wave lengths, does not differ enormously. If it did, double refraction would be obviously a colored phenomena, as is helical change of the plane of polarization and as is rotational magneto-optic change of the plane of polarization. These two last mentioned phenomena are entirely dispersive, and the amount of dispersion is more than four times as great for violet light as for red light. We shall come to that hereafter.

In double refraction therefore there is very little dispersion to consider, and μ^2-1 is very nearly constant. Writing this in the form $\frac{m_1 c_1 T^2}{1-\left(\frac{\kappa_1}{T}\right)^2}$ with a constant coefficient which I need not put down just now; we must have T considerably greater than $\varkappa_1$, so much greater that, writing this in the form $(m_1 - C_2 T^2)(1 + (\frac{\varkappa_1}{T})^2 + (\frac{\varkappa_1}{T})^4 + \ldots)$, the first term $(\frac{\varkappa_1}{T})^2$ will be sufficient to explain the dispersion. This gives a formula $(m_1 - C_2 \varkappa_1^2) - C_2 T^2 + (m_1 \varkappa_1^2 - C_2 \varkappa_1^4) \frac{1}{T^2}$ + another constant into $\frac{1}{T^4}$ &c, which agrees with Cauchy's formula because T is proportional to the wave length. It is quite certain that $C_2 T^2$ must be very small in comparison with m_1 in order that $\mu^2 - 1$ may be very nearly constant.

If we were to depend at all upon this term, $C_2 T^2$, for explaining the difference of refractive index in different directions we should have that difference directly proportional to the square of the wave length or four times as much for red as for violet light, which is not verified by observation. Not being able to help ourselves by that term, can we help ourselves in virtue of the appearance of C_2 in $\varkappa_1$? No, because C_2 is small in comparison with $\frac{\varkappa_1}{T}$. The only thing that might help us is difference of values of C_1 in different directions. That will give for difference of refractive indices,

$$\mu^2 - \mu'^2 = -\frac{C_1 - C_1'}{\rho} \cdot \frac{T^2 (C_2 - \frac{m_1}{T^2})^2}{(C_1 + C_2 - \frac{m_1}{T^2})(C_1' + C_2 - \frac{m_1}{T^2})}$$

Now can we in any way get anything constant out of that. Remark first that the factors of the denominator do not differ very much from our main denominator. Our main denominator expanded gives the comparatively exceedingly small change of value that corresponds to ordinary refraction, so that the denominator is approximately constant. Secondly, since $\frac{m_1}{T^2}$ is large in comparison with C_2 this difference is roughly $-\frac{C_1 - C_1'}{\rho D D'} \cdot \left(\frac{m_1}{T}\right)^2$. Thus the difference of our two refractive indices will be inversely proportional to the square of the wave length, and double refraction would be [illegible] a phenomenon as the effect of quartz upon polarized light producing the brilliant effects you know so well. This is absolutely out of the question for explaining double refraction.

I have been working in silence for a considerable time on this molecular theory. I became more and more interested in it and it has been a very great incentive to keep me at work upon it to have had the prospect of speaking upon the subject to you. I cannot but feel that there is a great reality in the theory of attached molecules. I cannot believe that the theory that does what it does in the way of explaining two or three of the phenomena that I have named, which have been the most enigmatical of all the phenomena of light according to the ordinary considerations, can be passed over; I cannot but believe that it is really true. But the explanation of double refraction remains ungiven by it.

I am able to explain the very finest lines that Rowland can show us, as well as the broad bands. Others have explained that so that I am only ambitious to point out what others have done in this direction: but what I wish to make noticeable is what others, I think, have not noticed so much viz: that we can do it without making away with energy. What seems to me important is to see how we can explain everything connected with observations of light by a definite communication of vibration to a system whose motions we can explain. As I have said two or three times before, the test of completeness and satisfactoriness in this kind of theory is, can we make a mechanical model of it. Take a perfectly elastic jelly, — that is a known thing, that we can have and look at and experiment upon. Fill it up with myriads and myriads of things like these molecular shells, and you can produce a solid which will transmit vibrations at a slower velocity than if the jelly were not modified by their presence; and if the rate of diminution of velocity thus produced follows somewhat nearly the law of the velocity of light in an ordinary medium, and if besides we can account for the energy that is not transmitted as waves in a particular case, with periods approximately so and so, like the case of sodium vapor, by showing that it exists in the molecules and that it reappears afterwards, and if we can account in that way for all variety of dispersions, and so on, then I say we can make something like a mechanical model illustrative of waves of light, so far as our theory is concerned.

I want to go somewhat into detail as to periods and magnitudes of masses and energies, so that there may be nothing indefinite in our ideas upon this part of the subject. I want, in the first place to call attention to two or three points connected with the possible density of the luminiferous ether. If any person present has seen a paper of mine, Note on the Possible Density of the Luminiferous ether and on the Mechanical Value of a cubic mile of Sunlight* I would be much obliged by him or her holding up a hand. I see Prof Forbes. No one else?

The very title of it is peculiar. In a reprint of it in a lithographed volume that was about ready to come out when I left England, I find a note of date Dec. 22, '82 to the following effect: "The brain wasting perversity of the Insular system which still condemns British Engineers to reckonings of miles and yards, and feet and inches and grains and pounds and ounces and acres is curiously illustrated by the title and num-

*Transactions of the Royal Society of Edinburgh 1854.

ical results of this article. The sacrifice of this Insular system that you heard discussed yesterday at the Congress would be made not only by us but Americans would make very much the same sacrifice. I believe engineers would save such an immense amount of labor in their calculations that in whole departments of drawing offices and designing offices in engineering establishments their occupation would be gone. The distinguishing feature of an engineer is the quickness with which he can reduce from square feet to acres, and so on. If his brain were free from that, he might do more elsewhere, and have more time to find out about the properties of matter. In illustration of this I have been here wasting brain on cubic miles and cubic feet instead of walking about and getting rested for this lecture. I am not going to go through that, however but I am going to try and make some estimate that you can understand, assuming that there must be a medium etc. I then thought that medium must be a continuation of our atmosphere. I would not say anything like that now.

"The first question that would naturally occur is, What is the density of the luminiferous ether in any part of space. I am not aware of any attempt having hitherto been made to answer this question; and the present state of science does not in fact afford sufficient data. It has, however, occurred to me that we may assign an inferior limit to the density of the luminiferous medium in interplanetary space by considering the mechanical value of sunlight as deduced in preceding communications to the Royal Society of Edinburg from Pouillet's data on solar radiation and Joule's mechanical equivalent of the thermal unit."

I want to ask in what proportion we may

increase the numbers that depend on Pouillet's estimate. I think it is 1½ or 1⅓. For instance 83 foot-pounds per second per square foot becomes not far from 100 foot-pounds per second per square foot, so that if the whole light and heat from the sun on a square foot is all absorbed, we have a heating effect corresponding to about 100 foot-pounds per second. That is a very definite experimental question. There are many doubts as to the accuracy of Pouillet's results, but not sufficient to shake them as being in the main a rough approximation to the truth. Many observers have repeated them and the tendency of observations since his time is to get larger and larger results. My impression is that Langley is inclined to reduce the figures. However, I am going to keep the figures as I have them here.

The mechanical value of a cubic mile of sunlight is 12000 foot-pounds, equivalent to the work of a one-horse power engine for one-third of a minute. There is something curious and interesting in that. The greatest volume of space lighted by the electric light is enormously short of the illuminating power of the sun over a cubic mile. It would be rather interesting to think of how many arc lights you must get into a certain space to have an initial illumination, let us say 1/100 of a horse power of work.

"This result may give some idea of the actual amount of mechanical energy of the luminiferous motions and forces within our own atmosphere. Merely to commence the illumination of three cubic miles, requires an amount of work equal to that of a horse power for a minute; the same amount of energy exists in that space as long as light continues to traverse it; and if the source of light be suddenly stopped, must be emitted from it before the illumination ceases. Similarly we find (the law of this being the inverse square of the distance) 15 000 horse power for a minute as the amount of work required to generate the energy existing in a cubic mile of light near the sun — 46,000 times as much as for a cubic mile of the sunlight at the earth's distance. The matter which possesses this energy is the luminiferous ether. If, then, we knew the velocities of the vibratory motions, we might ascertain the density of the luminiferous medium. or conversely, if we knew the density of the medium we might determine the average velocity of the moving particles. Without any such definite knowledge, we may assign a superior limit to the velocities, and deduce an inferior limit to the quantity of matter, by considering the nature of the motions which constitute waves of light. For it appears certain that the amplitudes of the vibrations constituting radiant heat and light must be but small fractions of the wave lengths, and that the greatest velocities of the vibrating particles must be very small in comparison with the velocity of propagation of the waves. Let us consider, for instance, plane polarized light, and let the greatest velocity of vibration be denoted by v; the distance to which a particle vibrates on each side of its position of equilibrium, by A; and the wave length, by λ. Then, if V denote the velocity of propagation of light or radiant heat, we have

$$\frac{v}{V} = 2\pi \frac{A}{\lambda};$$

and therefore if A be a small fraction of λ, v must also be a small fraction (2π times as great) of V. The same relation holds for circularly polarized light, since in the time during which a particle revolves once round in a circle of radius A, the wave has been propagated over a space equal to λ. Now the whole mechanical value of homogeneous plane polarized light in any infinitely small space containing only particles sensibly in the same phase of vibration, which consists entirely of potential energy at the instants when the particles are at rest at the extremities of their excursions, partly of potential, and partly of actual energy when they are moving to or from their positions of equilibrium, and wholly of actual energy when they are passing through these positions is of constant amount, and must therefore be at every instant equal to half the mass multiplied by the square of the velocity the particles have in the last mentioned case. But the velocity of any particle passing through its position of equilibrium is the greatest velocity of vibration, which has been denoted by v; and therefore, if ρ denote the quantity of vibrating matter contained in a certain space, a space of unit volume for instance, the whole mechanical value of all the energy, both actual and potential, of the disturbance within that space at any time is $\frac{1}{2}\rho v^2$. The mechanical energy of circularly polarized light at every instant is (as has been pointed out to me by Prof. Stokes) half actual energy of the revolving particles and half potential energy of the distortion kept up in the luminiferous medium; and therefore v being now taken to denote the constant velocity of motion of each particle, double the preceding expression gives the mechanical value of the whole disturbance in a unit of volume in the present case."

Actual energy was Rankine's word. The expression, kinetic energy, I am answerable for. I called that mechanical energy then. I had not begun to talk of kinematics as the science of motions and dynamics as the science of force, and I then used "mechanics" as it was generally used in books and universities and as it is sometimes used still.

"Hence it is clear — (Here is the point) — that for any elliptically polarized light the mechanical value of the disturbance in a unit of volume will be between $\frac{1}{2}\rho v^2$ and ρv^2, if v still denotes the greatest velocity of the vibrating particles. The mechanical value of the disturbance kept up by a number of coexisting series of waves of different periods, polarized in the same plane, is the sum of the mechanical values due to each homogeneous series separately, and the greatest velocity that can possibly be acquired by any vibrating particle is the sum of the separate velocities due to the different series. Exactly the same remark applies to coexistent series of circularly polarized wave of different periods. Hence the mechanical value is certainly less than half the mass multiplied into the square of the greatest velocity acquired by a particle, when the disturbance consists in the superposition of different series of plane polarized waves; and we may conclude, for every kind of radiation of light or heat except a series of homogeneous circularly polarized waves, that the mechanical value of the disturbance kept up in any space is less than the product of the mass into the square of the greatest velocity acquired by a vibrating particle in the varying phases of its motion. How much less in such a complex radiation as that of sunlight and heat we cannot tell, because we do not know how much the velocity of a particle may mount up, perhaps even to a considerable value in comparison with the velocity of propagation, at some instant by the superposition of different motions chancing to agree; but we may be sure that the product of the mass into the square of an ordinary maximum velocity, or of the mean of a great many successive maximum velocities of a vibrating particle, cannot exceed in any great ratio the true mechanical value of the disturbance. Recurring, however, to the definite expression for the mechanical value of the disturbance in the case of homomogeneous circularly polarized light, the only case in which the velocities of all particles are constant and the same, we may define the mean velocity of vibration in any case as such a velocity that the product of its square into the mass of the vibrating particles is equal to the whole mechanical value, in actual and potential energy, of the disturbance in a certain space traversed by it; and from all we know of the mechanical theory of undulations, it seems certain that this velocity must be a very small fraction of the velocity of propagation in the most intense light or radiant heat which is propagated according to known laws. Denoting this velocity for the case of sunlight at the earth's distance from the sun by v, and calling W the mass in pounds of any volume of the luminiferous ether, we have for the mechanical value of the disturbance in the same space, $\frac{W}{g}v^2$, where g is the number 32.2 measuring in absolute units of force the force of gravity on a pound. Now we found above $\frac{83}{V}$ for the mechanical value in foot-pounds of a cubic foot of sunlight; and therefore the mass in pounds of a cubic foot of the ether must be given by the equation $W = \frac{32.2 \times 83}{v^2 V}$. If we assume $v = \frac{1}{n} V$, this becomes

$$W = \frac{32.2 \times 83}{V^3} \times n^2 = \frac{32.2 \times 83}{(192,000 \times 5280)^3} \times n^2 = \frac{n^2}{3899 \times 10^{20}}$$

and for the mass in pounds of a cubic mile we have

$$\frac{32.2 \times 83}{(192000)^3} n^2 = \frac{n^2}{2649 \times 10^9}$$

It is quite impossible to fix a definite limit to the ratio $\frac{1}{n} = \frac{v}{V}$; but it appears improbable that it could be more, for instance, than $\frac{1}{50}$ for any kind of light following the observed laws. We may conclude that probably a cubic foot of the luminiferous ether in the space traversed by the earth contains not less than $\frac{1}{1560 \times 10^{17}}$ of a pound of matter, and a cubic mile not less than $\frac{1}{1060 \times 10^{6}}$."

The statement is not that these are the number of pounds of luminiferous ether in the cubic foot and mile, but that the number of pounds cannot be less than these figures, or else that the velocity of the vibrations will be more than $\frac{1}{50}$th of the velocity of light. Let us see what this ratio is. The corresponding statement as to amplitude and wave length would $2\pi \times$ amplitude $= \frac{1}{50} \times$ wave length, or amplitude $= \frac{1}{300} \times$ wave length. I think we can scarcely conceive of light coming away from the sun with vibrations through much greater amplitude than $\frac{1}{300}$ of the wave length. If it is not greater than that at the sun, then the mass of the luminiferous ether at the sun is 45000 times the number of pounds here given per cubic foot, or $\frac{3}{10^{16}}$ pounds, so that we may say that the luminiferous ether cannot contain less than this amount of matter in the neighborhood of the sun, and probably through the solar system. There are strong reasons for supposing that the density of the luminiferous ether is pretty nearly the same all through the solar system. In fact, all we know about the propagation of light seems to show that the refraction depends on the difference of effective density of the luminiferous ether and in so far as there is no sensible refraction, in all probability the luminiferous ether is very nearly of the same density.

I wish to make a little calculation, to show how much the luminiferous ether is condensed by the suns attraction. We are accustomed to call it imponderable. How do we know it is imponderable? If we had never dealt with air except by our senses, air would be imponderable to us. But we can, show that the weight of a column of air is sufficient to cause a difference of pressure on the two sides of a glass receiver. We have not the slightest reason to believe the luminiferous ether to be imponderable; it is just as likely to be attracted to the sun as air is. I do not like to make too many statements of that kind. At all events, the onus of proof rests with those who assert that it is imponderable. I think we shall have to modify our ideas of what gravitation is if we have a mass spreading through space with mutual gravitations between its parts without being attracted by other bodies. In the meantime, it is an interesting and definite question to think of what the weight of a column of luminiferous ether of infinite height resting on the sun will be, supposing the sun cold and quiet.

That is the same problem as that of the weight of the terrestrial atmosphere supposing it of equal density throughout. You all know the theorem for mean gravity in calling the energy at different distances inversely as the square of the distance. That applied to the case of one of the distances infinite gives the ordinary potential law. Take a column of height h and one square foot section resting on the surface of a body of the size of the sun (radius $= r$). The mean gravity will be $\frac{h}{r(h+r)} \times 28.6 \times r^2$, the density of the sun compared with terrestrial density being 28.6. Make $h = \infty$ and this becomes $28.6 \times r$ — I beg your pardon for going through all that; I ought to have known this result without finding it out. Unfortunately I only remember the sun's radius in miles from the woful defect of notation that is common to England and America. Call it 441000 miles or $.441 \times 10^6$. To reduce to feet multiply by 5280, and that by 28.6; and then that into the number of pounds in a cubic foot of the luminiferous ether. Will some of you kindly work that out. I make it 2×10^{-5}. * * * I am very glad to find that I am right, but I thought the possibilities were 100:1 that I was not.

I think we may say pretty safely that if the luminiferous ether is subject to gravity according to the same laws as are other bodies, the pressure per square foot on the suns surface—(setting aside the heat and motion of the sun) will be $\frac{2}{10^5}$ of the terrestrial weight of a pound. Compare that with the atmospheric pressure, which is 2000 pounds. We find $2000 \div \frac{2}{10^5} = 10^8$, so that the atmospheric pressure is one-hundred million times the ether pressure on the sun on the suppositions we have made.

Now, we have been supposing the luminiferous ether practically incompressible for light; but it does not follow at all that such a comparatively enormous pressure as $\frac{1}{50000}$ of a pound per square foot might not condense it. Of course this is very far beyond our knowledge. But if the luminiferous ether has the density indicated, the pressure certainly at the surface of a body like the sun would be one-hundred millionth of an atmosphere.

Lecture XVII.

I have written out a statement regarding Green's expression for the effect of Extraneous pressures. The formula for energy that Green gives on page 299 — not that Green called it that, he had not that name and merely called it a quadratic function — commences with the three terms which are written at the top of this paper, involving A, B, C. I have called this $2W$ for convenience. The other terms are those that we are familiar with for the case of symmetry but not farther reduced. I have not thought it necessary to write down more than the special terms I wanted to comment upon.

If you look at those terms, you see something quite unlike what appears in the equation of energy for an elastic solid as we know it. If we examine the meaning of those terms by taking our previous notation, a, b, c, for strains, and ϖ, ρ, σ for rotations we have the second set of formulae in this paper. What is meant here by rotations is not angular velocities as in the vortex motion theory, but angular turnings. For instance, the half of $\frac{d\eta}{dz} - \frac{d\zeta}{dy}$ expresses the angle through which the corresponding portion of the medium must be turned to bring it back to such a position that what it has experienced is merely an irrotational strain. In other words, if ξ, η, ζ be the actual displacements of any particle in the medium, viewed as functions of x, y, z, the dislocation of the material consisting

Fac-simile of Lecture Notes, Oct. 15th.

Green's expression for Effect of "Extraneous Pressures"

(p. 299)

$$2w = A\left\{\left(\frac{d\xi}{dx}\right)^2 + \left(\frac{d\xi}{dy}\right)^2 + \left(\frac{d\xi}{dz}\right)^2\right\}$$
$$+ B\left\{\left(\frac{d\eta}{dx}\right)^2 + \left(\frac{d\eta}{dy}\right)^2 + \left(\frac{d\eta}{dz}\right)^2\right\}$$
$$+ C\left\{\left(\frac{d\zeta}{dx}\right)^2 + \left(\frac{d\zeta}{dy}\right)^2 + \left(\frac{d\zeta}{dz}\right)^2\right\}$$

Put

$$a = \frac{d\eta}{dz} + \frac{d\zeta}{dy}\ ; \quad 2\varpi = \frac{d\eta}{dz} - \frac{d\zeta}{dy}$$
$$b = \frac{d\zeta}{dx} + \frac{d\xi}{dz}\ ; \quad 2\rho = \frac{d\zeta}{dx} - \frac{d\xi}{dx}$$
$$c = \frac{d\xi}{dy} + \frac{d\eta}{dx}\ ; \quad 2\sigma = \frac{d\xi}{dy} - \frac{d\eta}{dx}$$

We deduce

$$\frac{d\eta}{dz} = \tfrac{1}{2}a + \varpi, \quad \frac{d\zeta}{dy} = \tfrac{1}{2}a - \varpi$$
$$\frac{d\zeta}{dx} = \tfrac{1}{2}b + \rho, \quad \frac{d\xi}{dz} = \tfrac{1}{2}b - \rho$$
$$\frac{d\xi}{dy} = \tfrac{1}{2}c + \sigma, \quad \frac{d\eta}{dx} = \tfrac{1}{2}c - \sigma$$

Hence

$$2w = A\left(\frac{d\xi}{dx}\right)^2 + B\left(\frac{d\eta}{dy}\right)^2 + C\left(\frac{d\zeta}{dz}\right)^2$$
$$+ A\left\{\tfrac{1}{4}(c^2+b^2) + (c\sigma - b\rho) + \sigma^2 + \rho^2\right\}$$
$$+ B\left\{\tfrac{1}{4}(a^2+c^2) + (a\varpi - c\sigma) + \varpi^2 + \sigma^2\right\}$$
$$+ C\left\{\tfrac{1}{4}(b^2+a^2) + (b\rho - a\varpi) + \rho^2 + \varpi^2\right\}$$

I   II   III

in displacements of every parallel to the positions designated by ξ, η, ζ, whatever of strain it involves, involves a rotation through an angle equal to ϖ, ρ, σ.

Find $\frac{d\eta}{dz}$, etc., in terms of strain and rotation and we have the third set of formulas. Substitute these in the expression for 2 W and there results the last formula. Thus Green's formula, if it is true, implies that a certain amount of work would need to be obtained from the mere turning of each element, irrespective of the elastic forces between it and its neighbors. There is nothing that I can see in Green's assumption to correspond to that; there is no indication of any force that would produce it. The only way I see for producing anything of the kind would be by having two mediums mutually penetrating the space occupied and possessing some properties, of course not understood by us, according to which one of those mediums might resist the turning of the other relatively to it. But from the passage that I read to you yesterday from Green it is perfectly clear that he did not think of any thing of that kind. In the first place, as I said yesterday, the application of extraneous forces to a homogeneous isotropic solid cannot cause any difference in respect to the forces that would be produced by any dislocation superimposed upon that produced by the supposed extraneous forces— always provided the amount of the displacement is so small that the return force is simply proportional to the displacements of stresses represented by linear functions of the strains. If, however, this condition be not fulfilled, if stresses were applied so as to go beyond the proper limits of elasticity, or take first the case if there were a body that had proper elasticity through so wide a range that stresses might cease to augment in simple proportion to the strain and augment through more or less than a simple proportion, and if we were to apply extraneous forces to it sufficiently large to allow the deviation from simple proportionality to have any sensible effect, then A, B, C, terms such as those of Green would come into play. But under no circumstances that I can see could the rotational parts of Green's expression be true; and the only part of Green's expression that would have reality would be the first line, and the column marked III, in the last formula of this paper. But III observe, would correspond merely to a modification of the principal rigidities. In other words, that column may be written in the form $(B+C)\frac{1}{2}a^2+(C+A)\frac{1}{2}b^2+(A+B)\frac{1}{2}c^2$; so that it would be merely equivalent to adding $\frac{1}{2}(B+C)$.... to our rigidities (aa).... Also the first line is merely equivalent to adding A to our (cc), etc. I do not say, however, that we can adhere to Green's formula to this extent that when A, B, C, are the additions made to the direct tasinomic moduluses, then the additions to the rigidities would be $\frac{1}{2}(B+C)$, $\frac{1}{2}(C+A)$, $\frac{1}{2}(A+B)$.

Take the other case of the weakening effects of stress applied to a body beyond its limits of elasticity. By hammering, you develop in all probability aeolotropy in a body previously isotropic. You will see mentioned in my article on Elasticity an experimental proof of aeolotropy developed by such an action, showing the development of side long rigidities by torsion. A long straight steel piano-forte wire was twisted round through a great many turns — far beyond its limit of elasticity — and left to itself. Then it was found that when a weight was hung on it, it turned slowly in one direction and when the weight was taken off it turned back again. That

[NOTE]- Change above III into I and $\frac{1}{2}$ to $\frac{1}{4}$.

was proof of a development of aeolotropy in rigidity that made itself manifest in an obvious enough way by sidelong coefficients of rigidity. I do not feel that this expression of Green's goes towards expressing the physical theory of the introduction of aeolotropy in the properties of elastic solids such as is produced by hammering, with which we are all familiar etc. It would be interesting for physics if it were.*

[The terms II and III do not as I first thought express an impossibility. There is, in the case of an elastic medium subject to Green's "extraneous force," a dynamical relation to directions fixed with reference to the boundary of the portion of the medium concerned, which gives rise to return forces dependent on rotational displacements, analagous to the return force developed in a stretched cord by pulling it with equal force in opposite transverse directions, at the two points very near one another, so as to produce an infinitesimal rotation of the intervening portion. Then what is called Green's second theory (pp 305, 306 of his collected papers) does open a door for explaining the dependence of propagational velocity on direction of vibration instead of on the plane of distortion of the ether in a crystal. Stokes' explanation of this affair at the top of page 265 of his Report on Double Refraction (British Association, Cambridge, 1862), referred to, also on page 129 of his Burnett Lectures on Light (London, Macmillan, 1884), should be carefully read.]

I want to pursue a little further the dynamics of an elastic solid, especially with reference to the wave theory of light. Before going on to that, I turn to questions of aeolotropy. Weblike aeolotropy is, I believe, a very interesting and important

*Added Nov. 24th 1884.

subject in practical mechanics. The theory of it for a continuous elastic solid helps us in working out ideas that are important in respect to structures. In a structure as a whole, properties corresponding to to aeolotropy are produced in virtue of the manner of the structure. In fact, all structures of ironwork ties, and bracings, etc., are such that if we imagine a myriad of them put together—built up, as it were, like bricks—we should have an aeolotropic elastic solid. Our somewhat abstract questions of aeolotropy are closely connected with very important practical questions as to the mode of yielding of a body under the influence of certain definite forces. For example, take that of a tower made of diagonal bracings, etc., like that of the electric light tower to light the passage of Hell Gate in the harbor of New York. If any great weight is put upon the top of it, it will illustrate to us the same kind of sidelong aeolotropy in rigidity that the permanent twisting of a wire beyond its limits of elasticity develops in it. Generally the independent bracings of a tower are all placed symmetrically, so that nothing of the kind would happen, but take a tower braced unsymmetrically with diagonals all slanting one way, and there will be that kind of aeolotropy. I merely mention this as a somewhat crude illustration, just to show you that the theory of the continuous elastic solid is closely connected with subjects of great importance in engineering.

As to the physical properties of matter which are more properly subjects of interest and the subjects that we occupy ourselves with, I say it is an investigation of very considerable importance to find whether or not there is any of this weblike aeolotropy in crystals. Take crystals of the cubic class — crystals which have perfect equality and symmetry with respect to a cube. This is not a question of aeolotropy such as we have in the optical properties of biaxial crystals. The optical properties, as we have seen, are symmetrical with respect to the three axes. Our mechanical properties may or may not be so symmetrical. Take such crystals, which to appearance are absolutely similar in all their properties with reference to the six sides of a cube, and in reality seem to be absolutely similar in all physical properties as well, are they isotropic or not? There remains possible for them weblike asymmetry; and it seems not improbable that there will be weblike asymmetry of elasticity in cubic crystals. It may be very easily tested — or rather it is very easy to imagine a test. Think of what weblike asymmetry is with respect to a cube. It means more or less easier yielding to the distortion corresponding to a shear parallel to the faces than to the distortion corresponding to a shear parallel to the diagonals. Cut bars out in proper directions from such a crystal and test their flexural rigidity — that would be one way. This is not so easily done, however, because it is exceedingly difficult to get crystalline specimens, and to cut bars out of them. Other ways may be thought of. I merely speak of this thing to point out an interesting subject of research. Are there or are there not aeolotropic properties in respect to elasticity in crystals of the cubic class. We can make models as we have seen, of every kind of aeolotropy expressible by our 21 coefficients, and there is nothing easier than to make a model with weblike asymmetry. In fact, build up any structure with cubes

– build up a structure of packing boxes, and there is preeminently a structure with weblike asymmetry. Take a structure built of cubic bricks and the fact of there not being absolute continuity through the mortar gives to that structure most distinctly a weblike asymmetry. The elastic properties of solids are nearly related to the perfect elasticity developed in ideal, at least in connection with infinitely small displacements.

I do not need to put the question, is there any deviation from isotropy in crystals of the cubic class. The very first question of crystallography shows that there is. I remember a fine specimen of crystalline spar which Dr. Wm Cooper showed us quite 50 years ago, and knocked off a corner with a hammer. The fracture proved aeolotropy of strength. That well known elementary experiment shows us that the crystal is stronger in one direction than in another. That being so, does it not seem improbable that its moduluses of elasticity are all equal. It is a question of interest, and I had hoped to find ways of experimenting — I have not time to think of it now — and to experiment and find whether there were three moduluses of elasticity in crystals of the cubic class, and to get approximations to their magnitudes.

We have passed over preliminary considerations regarding double refraction. It is not necessary to spend any of the time that remains in going into the well known geometrical treatment of Fresnel's wave surface, whether we do it as Fresnel did it or get it from the elastic solid. That is sufficiently entered into in the various elementary works upon the subject. But

I want you distinctly to consider this question, What reasons have we for judging as to whether the direction of vibration is perpendicular to or is in the plane of polarization. To understand the meaning of the question, we must know what we mean by the plane of polarization. That is a mere technicality. The plane of incidence and reflection when light is polarized by reflection is called the plane of polarization. With that definition the question has a meaning. Of course, otherwise it might be confounded with the question how are you going to define the plane of polarization. I wish the question had come to us otherwise. I wish the plane of polarization had been defined in the beginning with respect to the vibrations and the question had been put more distinctly as a physical question, in respect to light polarized by reflection, viz. Is it when vibrations are in the plane of reflection and refraction that at a certain angle no light or but very little light is reflected, or is it when the vibrations are perpendicular to the plane of reflection and refraction that at a certain angle but little or no light is reflected. That is the physical question.

Mathematical literature has been loaded with a great deal of bad writing on this subject. A great number of investigations and statements called theories have been made, in which a piece of dynamical work is gone through; and then a condition is arbitrarily introduced; and that is called Cauchy's theory and something else is called Neumann's theory and something else is called MacCullagh's theory. I have perhaps done injustice in this statement to the possessors of these three names who have done such great things. I support myself however, in this statement by reading a few lines from Lord Rayleigh's paper on the Reflection of Light from Transparent matter. It is rather a celebrated thing. "Quite different from the foregoing is the theory of MacCullagh and Neumann, which is given in accessible form in Lloyd's 'Wave Theory of Light.' The following principles are laid down as the basis of investigation: — I. The vibrations of polarized light are <u>parallel</u> to the plane of polarization. II. The density of the ether is the same in all bodies as in <u>vacuo</u>. III. The <u>vis viva</u> is preserved; from which it follows that the masses of the ether put in motion multiplied by the squares of the amplitudes of vibration are the same before and after reflection. IV. The resultant of the vibrations is the same in the two media; and therefore in singly refracting media the refracted vibration is the resultant of the incident and reflected vibrations." One of these principles is simply an arbitrary assumption absolutely inconsistent with the dynamical conditions of the problem. If you want not to put too fine a point on it, you may call it MacCullagh's mistake or Neumann's mistake. Here is Lord Rayleigh's remark upon it: "When the vibrations are normal to the plane of incidence and therefore parallel in all three waves, the application of these principles gives rigorously Fresnel's tangent expression. If the vibrations are in the plane of incidence, the fourth principle alone leads to Fresnel's sine-formula. This only shows that the fourth principle is inconsistent with the others; for, as we shall see, unexceptionable reasoning founded on I and II leads to an altogether different result. The very particular case of IV required when the vibrations are normal to the plane of incidence happens to be correct."

Lord Rayleigh, I see, has the thing wrong so that I cannot show all the niceties of the wrongnesses of it. Everything about reflection and refraction of waves of light at the bounding surface separating two elastic solids is absolutely definite, and not hypothetical at all. Nobody can introduce a principle; it is a thing in which we have absolutely definite conditions to fulfil. I hope to put before you in a short form by to-morrow the conditions to be fulfilled and perhaps part of the work. You find the thing done absolutely correctly by Green, and you find Green's theory reproduced with some very important analytical improvement in the treatment of it by Lord Rayleigh, and in Lord Rayleigh's paper you find the thing worked out in a direction that Green left it unworked. Green, I may say, in a somewhat lax and not very well considered statement, assumes that the rigidities are equal in the two mediums and that the difference of wave length is due to difference of density. The only ?? it in this paper of Green's is manifest in what I read to you here: "The formula ~~which~~ which we have obtained are quite general and will apply to the ordinary elastic fluids by making $B = 0$ [That is rigidity = 0]; but for all the known gases A is independent of the nature of the gas, and consequently $A = A_1$. If, therefore, we suppose $B = B_1$ at least when we consider those phenomena only which depend merely on different states of the same medium, as is the case with light, our conditions become, etc." There is a note here: "Though for all known gases A is independent of the nature of the gas, perhaps it is extending the analogy rather too far to assume that in the luminiferous ether the constants A and B must always be independent of the state of the ether, as found in different refracting substances. However, since this hypothesis greatly simplifies the equations due to the surface of junction of the two media, and is itself the most simple that could be selected, it seemed natural first to deduce the consequences which follow from it before trying a more complicated one, and, as far as I have yet found, these consequences are in accordance with observed facts."

Now the analogy with gases is quite nonsense. I am rather surprised that Green put that in his paper as a reason for making $A = A_1$, because in his paper on the Reflection and Refraction of Sound he takes the reflection of sound at a surface of separation between air and water, in which the relation corresponding to this does not hold, and he points out how enormously far from holding is any such relation as this. I spoke of the disease of aphasia. This is a manifestation of it. What does one know of the meaning of A and B who can only speak of the properties of matter by "A" and "B". If Green had thought of the thing itself and not of the letters he would have saved himself that reference to gases at all. He would simply have said this, "Let us try the case of equal rigidities, and unequal densities," and he might have added, "This simplifies the formulae, and so far as I know, the results of the formulae with this simplification agree with observation." That is the state of the case. Everything else in Green is perfect. Lord Rayleigh improves the mathematical treatment by adopting that most valuable piece of shorthand, the imaginary symbol. Without the imaginary symbol, you have 8 equations in 8 unknown quantities. A skillful pilot will pilot himself among these 8 unknown quantities and pretty quickly find that they reduce to 4. But that is rather artificial work even for a skillful pilot among mathematical symbols. Lord Rayleigh's way can be followed by anybody acquainted with the mathematical forms and theorems he uses who is no pilot at all. The enormous value of this mathematical short hand — I owe that expression to Lord Rayleigh himself — is illustrated by no case better than by this. I do not care to use it when it does not help us; I prefer the sines and cosines; but when it saves ink and paper and brain let us by all means use shorthand.

Lord Rayleigh considers the question, Can you account for the known phenomena of the reflection of light, polarized and unpolarized, other than by supposing the rigidities equal in the two mediums and the densities unequal. He discusses the question penetratingly and by a particular test case he finds that it is impossible to get anything approximately of the same character as the real phenomena by the other extreme supposition which is admissible, that the difference of velocity in the two mediums depends on one of them being more rigid than the other, while their densities are equal. One of these suppositions, as Green found, gives results which somewhat approximately agree with the phenomena; the other, Lord Rayleigh proves, gives results exceedingly far from the phenomena.

Here is the state of the case: With the vibrations perpendicular to the plane of incidence in a wave of incident light, the supposition of equal rigidities and unequal densities in the mediums gives exactly Fresnel's law for light polarized in the plane of reflection. Alter this now by supposing the densities equal and the rigidities unequal and you get exactly Fresnel's formula for light polarized perpendicular to the plane of reflection. In other words the polarization of light by reflection could be accounted for by supposing the densities equal and rigidities unequal and the vibrations of polarized light in the plane of reflection, because in this case the light which is wholly transmitted and none of which is reflected consists in vibrations perpendicular to the plane of incidence. So far, therefore, we cannot judge between the two suppositions. But take the formula for vibrations in the plane of incidence. If the densities are unequal and the rigidities equal that gives us Fresnel's formulas. Those formulas are one of them rigorously, the other approximately the results of the full dynamical investigation corresponding to this supposition. But if we now take the other supposition, we get only one of Fresnel's formulas fulfilled, and the other excessively far from being fulfilled. It is absolutely impossible to get anything near to Fresnel's formula by supposing the vibrations of polarized light to be in the plane of incidence and reflection.

It remains to be considered whether by an intermediate supposition we can get any improvement in the result. For instance, suppose the density to be greater in one of the mediums and the rigidity to be greater but much less greater in proportion than the density. We might in that way get an improvement on the imperfect agreement for one of Fresnel's formulas without losing the perfect agreement for the other. But a full examination of that case leads to no satisfaction whatever.

We have an approximate agreement with Fresnel's formula on the supposition that the vibrations are perpendicular to the plane of incidence and that the action of the two media upon one another is that of homogeneous elastic solids; but the agreement is only approximate. Take Green's expression for the square of the ratio of the reflected and incident light $\frac{\beta^2}{\alpha^2} = \frac{K + (\mu^2-1)^4 \frac{b^2}{a^2}}{H + (\mu^2-1)^4 \frac{b^2}{a^2}}$. K vanishes at the polarizing angle; but what remains corresponds to a considerable deviation from zero in the amount of the reflected light. You find this gone into in an appendix to Green's paper on reflection and refraction of light. For the case of air and glass we find as much as $\frac{1}{49}$ for $\frac{B^2}{\alpha^2}$ at the polarizing angle. The amount of light reflected at the polarizing angle is very much less than that.

We cannot spend much more time upon this. Between Green and Lord Rayleigh we have the thing quite complete, so if I have explained it very badly to-day, you may make amendments to my explanations by reading Green and Lord Rayleigh. There are enough reasons here to make it very difficult to avoid the conclusion that the vibrations are perpendicular to the plane of polarization. But there are still stronger reasons than we have here. The strongest reason is of the kind first suggested by Prof. Stokes. It is closely related to his celebrated experiment on diffraction. I cannot say that it cannot be answered, but it seems to me that it is unanswerable. Good reasons for considering it unsatisfactory have certainly been given, but I think it probable that when the thing is fully examined it will be found that the conclusion may be still considered as rendered very probable, if not absolutely certain, by Stokes' diffraction experiment. But the experiment that seems most decisive is that on the polarization phenomena analagous to the blue of the sky. Stokes first suggested this, I believe, as a reason for supposing that the direction of vibration is perpendicular to the plane of polarization, but as Lord Rayleigh has shown, it was not so clear as Stokes supposed it to be. The view is this: Imagine a color to be produced by an enormous number of particles of diameters small in comparison with the wave length. The colors of the blue sky are only seen when the particles are known to be small in comparison with the wave length, which is not the case with colored dusts and halos, etc. Stokes view is that if the luminiferous ether is moving to and fro in the neighborhood of a particle the effect will be the same as if the ether were at rest and the particle moving, — the relative motion of the two being all that we have to consider. That being the case, it is obvious that the effect of a single spherule like that in the air, or of a vast number, would be to produce the kind of waves that we first considered. That is to say, waves with a zero motion in this direction and this ⟷ and oscillations to and fro perpendicular to the equatorial plane. You remember our formula with polarization in the equatorial plane. That is the kind of vibration we should have if the effect of the particles were as assumed in that view of Stokes. Therefore the light from a particle must consist of vibrations perpendicular to the plane which is perpendicular to the line through the center of the particle in the direction of the vibrations of the ether at the particle — the effect of the relative motion is that and cannot be anything else but that. Therefore all we have to do to find the direction of vibration in plane polarization is to test the polarization of light in the equatorial plane. The blue sky is complicated by the reflection from the surface of the earth, white clouds, etc. but in the main the light of the blue sky presents an almost complete polarization and a polarization in the plane through the sun — There is an almost complete polarization when we look in a direction at right angles to the direction of the sun. Experiments made on blue precipitates of various kinds all agree in this respect. Lord Rayleigh, however points out that there is another way of viewing the thing. We might in the first place assume that we have a dense mass, whose inertia prevents it from moving, but Lord

Rayleigh looks more particularly into the nature of the thing and considers this body as in many cases transparent. He considers the incidence of light upon it and passes continuously from the case of large drops of rain to the smaller drops of cloud white and the little particles of sodium or salt or spherules of dust or whatever they may be which cause the blue of the sky. He investigates fully the case when the particles are exceedingly small in comparison with the wave length. You must think of the light as reflected and refracted from the particle when it is large, and we are just brought back to the question I have put before you of the reflection of light at a transparent body. But when the particle is small in comparison with the wave length the theory of reflection and refraction at bounding surfaces does not at all follow. Lord Rayleigh works out the problem for equal rigidity and differing density and again for equal density and differing rigidity. The one is shown to come out exactly as Stokes pointed out but did not go into so fully which is represented here by the to and fro vibration. The other case is curious and is worth special consideration. I will put it down here and contrast it with the other. Suppose the spherule to differ from the rest of the medium in not having the same rigidity. What sort of vibrations will be produced. At the place of maximum displacement there is zero strain, but at the time when there is zero displacement there is a maximum of strain. Now when the difference is a difference of density this spherule will tell by its presence at the time when the acceleration of the medium to right or left is greatest, and the only effect on the medium is continuity of strain. On the other hand, if the densities are equal the motion of the ether will have no effect at the time of maximum acceleration and zero distortion;

it will have maximum effect at the time of maximum distortion. Let us put down an indication of the distortion of the luminiferous ether. We will have a slipping of one of these parallel lines with reference to the other. Suppose this sphericle has not the same rigidity as

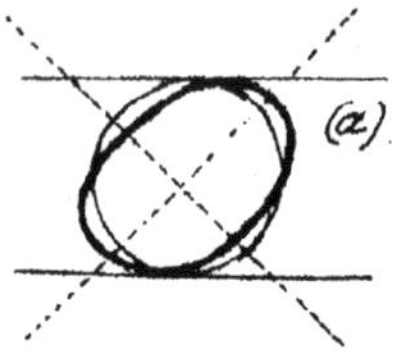
(a)

the luminiferous ether, it will be slewed from side to side in the manner I am indicating. It will be drawn out there (a) and in there. It is a bad drawing but it shows the principle. A circle is made oblique by sliding all the chords parallel to one diameter in one direction. A particle will then alternately be made oblique in this direction and made oblique in the other direction. I will put in dotted lines for the obliquities on either side. The vibration consists in an alternate elongation in a direction of 45° from the vertical on one side and an elongation in the direction of 45° on the other side of the vertical, with zero of change in the direction perpendicular to the board. Think of sphericles yielding to the distortion of the ether, but having more or less rigidity. That will cause them to act upon the medium in the same way as a vibrating body alternately getting longer and shorter in this direction and shorter and longer in this direction (dotted lines). That was one of our fundamental oscillations, our second case of motion, you will remember. There will be zero of effect in two lines at right angles to one another and maximum effect in directions perpendicular to those. Lord Rayleigh has pointed out that there will be no phenomenon corresponding to the zeros in this position. We may consider this test of Lord Rayleigh as settling the thing that Stokes overlooked. While Stokes says so and so, Lord Rayleigh says it is no so clear, but on looking into the thing, finds it must be so.

Thus we have absolutely proved that the direction of vibration is perpendicular to the plane of polarization, because we find that the plane of polarization, defined in the usual way and tested by Nicol's prism or what not, is the plane through the sun in the case of light reflected at right angles to the direction of illumination by a body consisting of minute spherules separate from the luminiferous ether. I shall try to put a little more clearly on paper the state of the case in reflection and refraction to-morrow. I had intended to say something upon molecular dynamics to-day, but alas, the time has all gone. I have used my opportunities very imperfectly in bringing this subject before you, but we must make the best of it, notwithstanding.

Lecture XVIII.

I have tried to put down something regarding the reflection and refraction of waves at the surface separating two homogeneous mediums, vibrations to be in the plane of the three rays. I took that case at once because the other cases are so exceedingly easy that it does not matter much whether we take them or not. You will find them thoroughly and simply worked out in Green and also Lord Rayleigh's

Lecture Notes of Oct. 16.

Refraction and Reflection at Interface between two Homogeneous Elastic Mediums.

I – Vibrations in the plane of the three rays – This Plane xy.

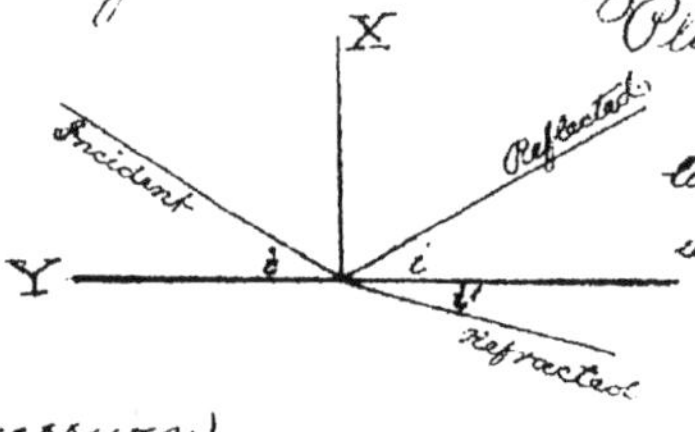

Lines indicate wave fronts

$$\xi = \frac{d\varphi}{dx} + \frac{d\psi}{dy}$$
$$\eta = \frac{d\varphi}{dy} - \frac{d\psi}{dx} \quad \Bigg\} \ldots\ldots (1) \qquad \delta = \nabla^2\varphi$$

$(k - \frac{2}{3}n)\,\delta = p \quad \ldots\ldots (2)$

($-p$ corresponds to fluid pressure.)

$$\left.\begin{array}{l} P = p + 2n\frac{d\xi}{dx}, \quad Q = p + 2n\frac{d\eta}{dy}, \quad R = 0 \\ S = 0,\ T = 0,\ U = n\left(\frac{d\xi}{dy} + \frac{d\eta}{dx}\right) = n\left(2\frac{d^2\varphi}{dxdy} + \frac{d^2\psi}{dy^2} - \frac{d^2\psi}{dx^2}\right) \end{array}\right\} \ldots (3)$$

T & T' § 697

$$\rho\frac{d^2\xi}{dt^2} = \frac{dP}{dx} + \frac{dU}{dy}, \quad \rho\frac{d^2\eta}{dt^2} = \frac{dQ}{dy} + \frac{dU}{dx} \quad \ldots\ldots (5)$$

These without accents refer to upper medium with " lower "

$$\left.\begin{array}{l} \psi = A\varepsilon^{\iota(ax+by+\omega t)} + A_{,}\varepsilon^{\iota(-ax+by+\omega t)} \\ \varphi = B\varepsilon^{-\bar{b}x+\iota(by+\omega t)} \\ \psi' = \quad \varepsilon^{\iota(a'x+by+\omega t)} \\ \varphi' = B\varepsilon^{\bar{b}x+\iota(by+\omega t)} \text{ where } \omega = \frac{2\pi}{T}. \end{array}\right\} \ldots\ldots (6)$$

By (1), (2), (3) and (5) we have
$$\left.\begin{array}{l} \rho\frac{d^2\psi}{dt^2} = n\nabla^2\psi \\ \rho\frac{d^2\varphi}{dt^2} = (k + \frac{4}{3}n)\nabla^2\varphi, \end{array}\right\} \ldots (7)$$

From (6) and (7) we find $b^2 - \bar{b}^2 = \rho\omega^2/(k + \frac{4}{3}n) \quad \ldots\ldots (8)$

$\rho\omega^2 = n(a^2+b^2) = n(a'^2+b^2) \ldots (9)$

At interface ($x = 0$) we have

$$\left.\begin{array}{l} \xi = \xi',\ \eta = \eta', \text{ by continuity of matter} \\ \text{and } P = P',\ U = U', \text{ by balance of forces} \end{array}\right\} \ldots (10).$$

We have, by (8), and (2), and (8) and (9), The part of P dependent on $\varphi_s = \{(k - \frac{2}{3}n)(\bar{b}^2 - b^2) + 2n\bar{b}^2\}B = \{-\rho\omega^2 - n(\bar{b}^2 - b^2) + 2n\bar{b}^2\}B = n\{-(a^2+b^2) + (\bar{b}^2 + b^2)\}B = n(\bar{b}^2 - a^2)B \ldots (11)$

Hence, and from (6), and (1)

$$\left.\begin{array}{lll} \xi=\xi' & \text{gives} & -\mathfrak{b}B+ib(A-A_{,}) = \mathfrak{b}B'+ib \\ \eta=\eta' & " & ibB-ia(A-A_{,}) = ibB-ia' \end{array}\right\} \quad \dots\dots \quad (12)$$

$$\left.\begin{array}{lll} P=P' & " & n\{(\mathfrak{b}^2-a^2)B-2ab(A-A')\} = n'\{(b^2-a'^2)B'-2a'b\} \\ U=U' & " & n\{-2ib^2B+(a^2-b^2)(A+A_{,})\} = n'(2b^2B'+a'^2-b^2) \end{array}\right\} \quad \dots \quad (13)$$

For case of sound don't use (11); but early take $n=0$. Results Green, quite simple and perfect.

For case of incompressibility take

$$\mathfrak{b} = b \quad \dots\dots\dots\dots \quad (13$$

For general supposition of finite compressibility $\mathfrak{b}$ must, by (8), be either wholly real, or wholly imaginary.

Case I is too small to allow condensational wave to be generated by the reflection and refraction:

Case II., $\rho\omega^2/(k+\frac{4}{3}n) > b^2$;

a condensational wave is generated. If l be its wave-length

$$\frac{2\pi}{l} = \sqrt{(b^2-\mathfrak{b}^2)} = \sqrt{\frac{\rho\omega^2}{k+\frac{4}{3}n}} \quad \dots\dots\dots \quad (14)$$

as we knew long ago.

I will say a few words in explanation of the different formula. In the first place, we have motion in two dimensions alone, and our formula belong therefore to the general formula with that limitation to two dimensions and coördinates x, y — z not appearing. In our original division of the solution into a condensational part and a distortional part, Φ in equation (1) corresponds to the first, and Ψ to the second; for observe that Ψ expresses here a solution for which $\frac{d\xi}{dx}+\frac{d\eta}{dy}=0$, which is the condition of no dilatation. We have separated the solution therefore, merely by a functional device, into these two parts. As we are going to apply these solutions to the case in which the medium is incompressible, so that ~~the~~ condensation is impossible, I will introduce a new word. Insted of "condensational wave", we will talk of "pressural wave;" and we shd find that at the bounding surface of a medium we have a pressural wave, even if the medium be incompressible. I have brought in $p = (k - \frac{2}{3}n)\delta$ because it does not become infinite at all when k becomes infinite, the other factor δ becoming zero. Verify the value of P and Q which appear in equations (8) by substituting for p the value $(k - \frac{2}{3}n)\delta$ and you will find that they come to forms written out for them in one of the earlier lectures. I put down a form for P you may remember that I said was convenient for some purposes (p. 25). This being a case of motion in two dimensions, the shear is wholly in the plane xy. $\therefore R = 0, S = 0, T = 0.$ The value of U obtained from its fundamental form completes equations (8). In point of fact and as matter of arrangement, I need not have written down the general equations of motion (5) at all, but might simply have taken equations (7) from our old friends the differential equations (on p. 33) which φ and ψ must fulfil. However, it is well to put it down from the beginning and to verify for yourselves that equations (7) are derivable from (1), (5), etc. All these formulas used without accents refer to the upper medium; all with accents refer to the lower medium. I use the words upper and lower merely for convenience and as corresponding to this diagram, without reference to the actual positions of the mediums. We might have, for instance, a case in which water was the upper medium of our diagram and the lower medium air. This is a case in which the introduction of the analytical shorthand is very valuable. Try this case without it, and you will find you have 6 equations in 6 unknown quantities. The analytical shorthand reduces the problem to 4 unknown quantities, A, A_1, B, B'. I use the symbol ι for $\sqrt{-1}$, i, i' being reserved for the angles of incidence and refraction.

ω is the angular velocity of the relatively circular motion, or $\omega = \frac{2\pi}{T}$. The object is to express a simple harmonic motion. The advantage of the mathematical shorthand consists in the fact that a similar set of formula holds for $-i$ as well as for i. You can realize by adding, obtaining sine and cosine formula. Just remark the term $\varphi = B\varepsilon^{-bx+i(by+\omega t)}$. If B comes out in our result a real quantity change the sign of i and add. That gives a cosine. If B comes out a pure imaginary, change i into $-i$ and subtract. That gives a sine! In reality B will come out mixed real and imaginary, and there will result this form, $\varepsilon^{-bx}(C\cos(by+\omega t) + D\sin(by+\omega t))$ I have taken out this term because we want to look a little more particularly at it.

Let us think of the meaning of these different terms. There is only one plane distortional wave in the lower medium, because by hypothesis the light is incident in the upper medium upon the separating interface. We must then have in the lower medium an expression for a refracted plane wave, and, if we cannot get quit of it, we must have a pressural wave. That is then what is denoted by ψ' and φ'. For the sake of symmetry I have chosen the refracted distortional wave as being the given one of the set. That is why it appears with no second coefficient; and also not wanting to use more coefficients than necessary, I let the single coefficient of ψ' be unity. The remaining coefficients bring in the 4 unknown quantities. There is something more to be said as to what is known and unknown. What of the $a, b, \mathfrak{b}, \omega$? We shall suppose ω to be known, and the moduluses to be known. The equations of motion then give us the $a, b, \mathfrak{b}$ as in equations (8), (9). In strict analytical propriety we must not know the x, y, t, in the second medium;

but we do the accented a's and b's. We accent a because it is clearly different in the two mediums. We do not accent the b because it has clearly the same value in the two mediums. Put down the values of a and b in terms of the wave length and then the thing will be perfectly clear. Let λ be the wave length of the plane distortional wave in the upper medium; we have then in the upper medium $-(ax+by) = \frac{2\pi}{\lambda} r$, r being the perpendicular distance from the focus to the wave front. Since i is the inclination of the wave front to the axis of y, we have $a = \frac{2\pi}{\lambda} \cos i$, $b = \frac{2\pi}{\lambda} \sin i$. For perpendicular incidence we have $i=0$, giving the augment $\frac{2\pi x}{\lambda} + \omega t$ which is correct. For grazing incidence, $i = 90°$, giving $\frac{2\pi}{\lambda}y + \omega t$, which is correct. We may treat similarly the pressural wave, in the cases in which it extends into the second medium as a plane wave, letting l be the wave length as in the paper. You might just add the above to the paper in explantion of the notation a, b, in equations (6). It is so difficult to write with the jelly-graph ink that I economized as much as possible.

At the interface, that is to say, the position $x=0$, we have continuity of matter. Hence $\xi = \xi'$, $\eta = \eta'$. Again, we have nothing to do with Q at the interface between two mediums, because Q is a force that acts on the surface perpendicular to the interface. If we consider the forces on the interface, there must be a balance between them. Therefore $P = P'$, $U = U'$. These are the conditions to be solved. That leaves a clean simple problem of dynamics, and yet people have been working at it for 50 years and have left it in a very sadly muddled condition, with the exception of the clear accurate, and very comprehensive papers of Green and Rayleigh. The thing that has introduced the difficulty, and makes this a more complicated difficulty than the other cases is the pressural wave. The pressural wave, in fact, has been the *bete noir* of this problem. I do not know how Cauchy treats the animal. Somehow, he introduces fallacious terms involving consumption of energy. MacCullagh and Neumann killed the animal with bad treatment. Sam Haughton yoked it to an Irish Car and it would not go. Green and Rayleigh have treated it according to its merits and it has escaped whipping at their hands.

There is a little novelty in this way of treating it expressed in (11). I have got the thing into a form in which I avoid the question of compressibility or incompressibility until we are supposed to take it up definitely. In equations (11), I want to get the part of P dependent on Φ. That is the thing in which a little management is required to avoid difficulties. It works out rigorously from the preceding fundamental formula. Note the two little b's and distinguish between them for the present. In the first modification we introduce $\rho\omega^2$, by (8). In the next modification we get rid of $\rho\omega^2$ by using that of the two values of (9) which belongs to the upper medium, viz: $\rho\omega^2 = n(a^2+b^2)$. Thus the part of P depending on Φ takes the simple form $n(b^2-a^2)B$.

I have worked this problem out in this paper more fully than has been done in Lord Rayleigh's paper. It would take too long to go all through it for you. I have done it at various times, chiefly in steamers and on railways. I came in that way quite unexpectedly upon this result — I am not going to give much time to it though, because it is not really of importance for light — I found a very curious expression which gives us a case of complete polarization. At the regular polarizing angle, the following relation involving unequal rigidities, $na(a-3a') = n'a'(a'-3a)$ brings about a vanishing of the imaginary part of A, and therefore leads to a case of complete polarization by reflection. The a, a' must have the values corresponding to the angle of polarization, which is the same as Fresnel's angle. For n = n' the result is null. This relation implies a greater rigidity in the medium of slower wave velocity and as the medium of slower wave velocity has a greater refractive index, it implies greater density also, — but not greater in the same proportion as the rigidity. Now I have looked at the light that would be reflected at direct incidence and find that it is very much in excess of what would be given by this. The ratio of the intensity of reflected to refracted ray for direct incidence on the supposition of equal rigidities and unequal densities is $\left(\frac{\mu-1}{\mu+1}\right)^2$. For the case of glass take $\mu = 1.5$ and that becomes $\frac{1}{25}$. I tried, and as nearly as my rude experiment allowed me to judge, something like a tenth part of the light was reflected from a piece of ordinary glass. The whole light reflected from two surfaces should be approximately double that from one. Thus my own rough experiments showed that Fresnel's formula was so nearly correct that I was quite unlikely to make anything out of this supposition of unequal rigidities. From that moment the algebra lost its interest for me. I shall put it in form sometime or other; whether in time to be incorporated in the report of these lectures or not is not of great consequence to you. I just tell you about it, however. It is worth knowing that a thing may be examined in this way and that way, and what sort of possibilities there are in it. I would not altogether discard the possibility of the rigidity being different in the two mediums for all cases. Our knowledge of transparent bodies is, in fact, very limited, and that knowledge is confined chiefly to visible light. When we investigate these things for invisible chemical light, and for dull radiant heat, we may find something very different from what we at present suppose to be the state of things as regards the answers to these fundamental questions. I note, indeed that the reflection of radiant heat from rock salt seems to be much greater than according to Fresnel's formula. Green takes $n = n'$ because it simplifies the work. Lord Rayleigh supposes the rigidities to be equal and unequal densities to be the cause of difference of velocity. In his paper on the reflection of light from small particles his reasoning is very urgent and seems exceedingly binding on this subject; we can scarcely get away from the conclusion that the rigidities are equal or very nearly equal and the difference of velocity does depend on on difference of density. He shows that if we make any considerable deviation from the position of equal rigidities we induce effects not known to observation and not known to observation. Most particularly telling is the polarization of light from fine particles. By Lord Rayleigh's work it seems that if there be any sufficient difference of rigidities to be worth thinking of in the way of explaining outstanding difficulties of another kind, the polarization that we have will be annulled and we shall not have nearly a good enough approximation to the polarization to represent the state of the case.

That being the case, the question is left, what can we make of the results of these equations. The results are given in Lord Rayleigh and Green. I unfortunately, yesterday, did not come upon the right paper; I will call your attention to it once more because I want to speak of magnitudes. I want to show you that we are very far indeed from an agreement with observation in the formula derived from these processes. We ought to find from these processes that our reflected light very nearly vanishes at a certain angle of incidence. Green works it out and gives a formula. The actual minimum value of that formula is not quite that which

Green gives, but in an appendix by Ferrers the true value is given. For the case of air and water, $\mu = \frac{4}{3}$, and Green finds for the minimum value of the intensity of the reflected light, $\frac{1}{151}$. Now compare that with the light reflected from water by direct incidence. By Fresnel's formula derived by the same mathematics, that is $(\frac{4}{3}-1)^2/(\frac{4}{3}+1)^2 = \frac{1}{49}$. How could Green say that his result was, as nearly as he knew, conformable with observation, when he finds that the light at the polarizing angle is a third part of that reflected by direct incidence. It is nothing like a third part. Speaking roughly, I do not believe the light reflected at the polarizing angle is a 20th part from the nullness of the amount of light that is left at the polarizing angle when you apply the light in the usual way. Try it and you find that proportion is enormously less than the proportion Green gives. Ferrers helps a little out of the matter by saying that instead of Green's minimum value of $\frac{1}{151}$ we have more accurately $\frac{1}{166}$; but that is at an angle not quite agreeing with Fresnel's polarizing angle, which does not make matters much better. It is moreover so small an approach to the annulment of light that we have that it cannot show anything satisfactory. Take the case of glass ($\mu = 1.5$) in which the intensity of the reflected light at the polarizing angle is $\frac{1}{47}$ and for direct incidence, it is $\frac{1}{25}$. Actually in the case of glass there is not at the polarizing angle, anything like half the light at direct coincidence. The formula is simply a failure. Green did not notice this; he had switched off on something else, I dare say. To be sure $\frac{1}{151}$ is a small number and it looks as if it might be right, but if he had considered how small the reflection really is, he would have seen that that is no approach to a satisfactory explanation. I will just give you the formulas, because some of you may not have access to Lord Rayleigh's papers [Phil. Mag. Aug. 1871]* The ratios of the amplitudes of the reflected and incident vibrations is given by

$$\frac{R'^2}{R^2} = \frac{\cot^2(i+i') + M^2}{\cot^2(i-i') + M^2}, \text{ where } M = \frac{\mu^2-1}{\mu^2+1}.$$

It may be interesting for you to work that out for yourselves, which you can do from our equations.

Besides the minimum ratio attained when we vary the direction of the incident light from normal to grazing incidence, there is a change of phase. If we had complete polarization the state of things would be this: phase remaining perhaps constant until the intensity diminishes to zero, then the phase changing suddenly as the inclination passes through the zero position. What really happens according to the formula is: phase varies gradually; at the minimum; ~~and at the polarizing angle~~ it is, roughly speaking, midway between the phase corresponding to direct incidence and the phase corresponding to grazing incidence. The want of complete fulfilment is connected with the gradual change of phase. In observations we can only take the relative phase – the difference of phase between the two component rays, i.e. the component consisting of vibrations in the plane and perpendicular to the plane. Lord Rayleigh refers to Jamin here and says, "now what is observed in experiments is the acceleration or retardation of one polarized component with regard to the other and is therefore given simply by difference between the two angles. The ambiguity must be removed by the consideration that when the incidence is normal, there is no relative change of phase, though throughout Jamins papers it is assumed that there is in that case a phase difference of half a period. I am at a loss to understand how Jamin could have entertained such a view, which is inconsistent with

* Other papers of Lord Rayleigh's referred to are in Phil. Mag. Feb. April & June 1871, May 1872.

continuity, inasmuch as when $i=0$ the distinction between polarization in the plane of reflection and polarization in the perpendicular plane disappears."

In this paper I have only given you the reflection and refraction for the case of vibrations in the plane of the three rays. The case is so exceedingly simple for vibrations perpendicular to the plane of the diagram that you will not regret my not having given it to you. It brings out exceedingly simple formulas which agrees exactly with Fresnel's sine formula when we suppose the rigidities equal and the densities unequal. Very curiously, densities equal and rigidities unequal, give you Fresnel's tangent formulas; and it gives you complete polarization — that is a most interesting result. What is more, it gives you the same intensity for light reflected at direct incidence as Fresnel's formula. You might think that would be a good foundation for allowing that the vibrations were in the plane of polarization. But alas for that supposition, Lord Rayleigh has shown that it is absolutely impracticable in the problem of vibrations in the plane of the three rays to get anything approaching to Fresnel's formula at all, if you take the densities equal and the rigidities unequal.

We cannot but conclude from all we have before us, that the theory of the homogeneous elastic solid is quite unsatisfactory in respect to polarization, the approximation to explanation of the extinction of the ray consisting of vibrations in the plane of the three rays being so exceedingly, so monstrously rude as we have seen. I am surprised that it has not been denounced more by others who have touched upon the subject.

I would like to call your attention to Green's formula for refraction of sound. You have got the formulas down here passing over (13), and that beautiful result of Fresnel becomes exceedingly simple. The ratios of the intensities or the squares of the ratios of the displacements is,

$$\left(\frac{\rho'}{\rho} - \frac{\cot i'}{\cot i}\right)^2 \Big/ \left(\frac{\rho'}{\rho} + \frac{\cot i'}{\cot i}\right)^2$$

For the case of all gases, $\frac{\rho'}{\rho} = \frac{\sin^2 i}{\sin^2 i'}$ and the above formula reduces to $\frac{\tan(i-i')}{\tan(i+i')}$ or Fresnel's tangent-formula. There then is an agreement with one Fresnels most remarkable formulas for sound reflected at an interface of separation between two gases of different densities. On the other hand, if we have anything like the law of relation between bulk moduluses on the one hand and densities on the other that we have between air and water, or between two different liquids, we have no approach to this formula. It is not easy to see how that formula for sound can be verified by experiment, but still the result is in itself exceedingly interesting.

For the case of incompressibility, we must take $\mathfrak{b} = \mathfrak{b}$. That gives us our relation $\delta = \nabla^2\varphi = 0$. It is interesting to remark that without taking $\mathfrak{b} = \mathfrak{b}$ we have a set of formulas that may be used. In some cases those formulas will give condensational waves; in others not. Instead of saying what is under case I you might cancel it, and say according as $\frac{\rho\omega^2}{k+\frac{4}{3}n}$ is less than or greater than $\mathfrak{b}^2$ we have case I or case II. You will see then that inasmuch as $\mathfrak{b} = 0$ for direct incidence, ~~So~~ that for incidences not too oblique we always have condensational waves: for very oblique waves we have no condensational waves. For direct incidence the condensational wave, as you may easily see from working out the formula, is null; it is necessarily null. But for incidence nearly direct, the condensational wave is not null and it can only be annulled by making $k = \infty$.

With respect to my very faulty expression

regarding Sam Haughton having yoked his animal to an Irish car, I meant to say that he tried to make this condensational wave help the car out of the ditch in which it is lodged — that is to say, he tried to get us out of our difficulty by aid of the difference between b and b'; but it would not work.

We will put this condition for the existence or non-existence of a condensational wave in a better form. We have $b = \frac{2\pi}{\lambda} \sin i$, where λ is the wave length of the distortional wave. The relation between λ and ω is as follows: v (the velocity of propagation of the distortional wave) $= \frac{\omega\lambda}{2\pi} = \sqrt{\frac{n}{\rho}}$ $\therefore \rho\omega^2 = \frac{4\pi^2 n}{\lambda^2}$. These values substituted in $\frac{\rho\omega^2}{k+\frac{4}{3}n} - b^2$ give $\frac{4\pi^2}{\lambda^2}\left\{\frac{n}{k+\frac{4}{3}n} - \sin^2 i\right\}$; $\therefore$ the equations for finding the critical angle is $\sin^2 i = \frac{n}{k+\frac{4}{3}n}$. We have the conclusion that if the angle of incidence is anything less than that given by this formula, there is a condensational wave, unless the angle is zero — then we have no condensational wave. And if i is greater than that critical angle, there is no condensational wave. That I think absolutely settles the whole question with regard to the condensational wave.

There are two or three things that I wish to speak about. I want to clear off at once the question of helicalness on the plane of polarization, commonly called the non-magnetic rotary effect. I have objected to the name rotary, because it is not properly applied, and have taken the name helical because the phenomenon essentially depends on a screw like form somehow or other. So far as I know, the first place where this distinction is pointed out and the essential connection of the Faraday property with rotation shown is in a paper of my own in the Proceedings of the Royal Society of London, May 1856. I just read two or three sentences from that paper: "The elastic action of a homogeneous strained solid has a character essentially devoid of all helical and of all dipolar asymmetry. Hence the rotation of the plane of polarization of light passing through bodies which either intrinsically possess the helical property (Syrup, oil of turpentine, quartz crystals &c.) or which have the magnetic property induced in them, must be due to elastic reactions depending on the heterogeneousness of the strain through the space of a wave or to some heterogeneousness of the luminous waves," etc. But here is the point to which I wish to come. "Imagine for example a liquid filled homogeneously with spiral fibers or a solid with spiral passages through it of steps — I said here — of not less than forty millionth of an inch — meaning steps of a screw not less than a thousandth of the wave length. This "might be certainly expected to cause either a right handed or a left handed rotation of ordinary light." There can be no doubt that this is the correct explanation. For a rough mechanical model of a medium possessing helical properties, take a jelly and bore ever so many cork screw holes in it — that will introduce a heterogeneousness of structure with a definite spiral character. Take another jelly and bore it with left handed cork screw holes and that will induce a definite spiral structure also. One of those mediums seen in a looking glass would look like the other; we have that kind of want of symmetry that there is between the right and left hand. Another example is to take a bunch of spiral springs and fill up the interstices with mortar, jelly or something of that kind, and you will have that property. If the wave length be enormously great in comparison with the dimensions of heterogeneousness, the turning effect on the plane of polarization will be exceedingly small. It will be null if the wave length is infinite in comparison with the dimensions corresponding to the heterogeneousness. It seems almost certain that this worked out would give us, I will call it, the rotary effects (although I protest against the name) of quartz, etc, somewhat nearly according to the well known formula of inversely as the square of the wave length. You know that in reality the practically constant quantity, square of the wave length into the amount of the rotation, does gradually increase as the frequency increases for the substances that have been experimented on, so that the rotation varies more than according to the inverse square of the wave length. I see it is stated that Biot has worked this out. If he has worked it out right, it is exceedingly interesting and important. The other question of the Magnetic influence on light I shall say nothing about.

I had hoped to bring forward an addition to our molecular theory, showing you definitely the rotation of the plane of polarization produced by introducing an enormous number of gyrostats into our jelly. I will show you how the thing may be done, and I will tell you why I do not give the mathematics of it. One reason is that to-morrow is our last day, otherwise I would try to give you the mathematics of it, unsatisfactory as it is.

Suppose we have here distortional waves. The arrow heads indicate the to and fro motion of a wave in the plane in question. Besides the distortion there will be rotation. Suppose we have our massless rigid shell lining or spherical cavity in the ether; and on that lining let us pivot by a proper shaft a fly wheel like the fly wheel of a gyroscope and suppose that to be rotating with enormous rigidity. The jelly may move this way or that way without inclining the axis of that fly wheel. But force the axis to turn in the plane of the board, and that introduces a tort pressing upon the bearings of the ends of the fly wheel in

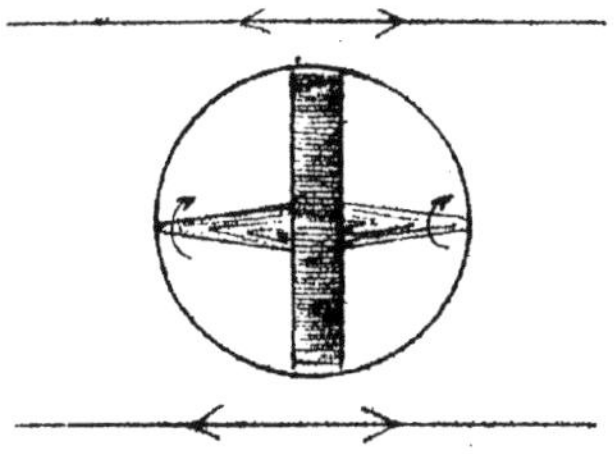

a plane perpendicular to the board. The plane of that sort is, of course, the plane of the axis perpendicular to the plane of our diagram. That transverse force is very easily introduced into the equations of motion and it gives us just what we want if we only want to show rotation of the plane of polarization. It gives you rotation of the plane of polarization following Faradays law that if you send the light in one direction you get a rotation of the plane of polarization. Send it backwards or forwards in the direction of the revolution of the plane of polarization and it goes on rotating, as you all know. That is satisfactorily explained by this gyrostat — nothing could be more satisfactory or clear than it is.

In the time that I have been talking about it, I might have put down the symbols. Why do I not go into it, and try to be make it a part of our molecular dynamics? I answer because I cannot bring out the law of inverse proportionality to the square of the wave length, which observation shows to be somewhat approximately the law of the phenomenon. If you deal with it in this simple way, it comes out inversely as the wave length and not inversely as the square of the wave length. Until a week ago, I thought that by putting a fly wheel somehow or other into our molecule I could get a rotary effect according to which the magnitude would vary according to two terms, $\frac{C}{\lambda} + \frac{C'}{\lambda^3}$. If that were so, I could bring the thing to vary according to observation, because there is no rigorous agreement to the inverse square of the wave length. it varies more than that and it is possible that it will be expressed by some such formula

But alas, my results give me the other law, not more effect with greater frequency, but less effect with greater frequency, "according to the inverse wave length." I therefore lay it aside for the present, but with perfect faith that the principle of explanation of the thing is there. I cannot pretend that the very simple matter of molecular dynamics at which I am driving has accomplished the solution of any great difficulties, but I do think it is of high importance and interest.

*[Referring to rotational, or Faraday-magneto-optic, effect: when I said, "I get persistently $\frac{1}{\lambda}$ for the law, but it is to be $\frac{1}{\lambda^2}$ approximately but varies a little more than that, etc." I was under a misapprehension, to be now corrected as follows:

This result I have found is that circularly polarized light travels with different velocities according as the orbital motions are with or against the Amperian currents: this difference of the two velocities being, for lights of different homogeneous colors, directly as the frequency of the vibrations. The resultant of two circular motions of equal periods, in opposite directions in the same circle is simple harmonic vibration along a diameter in the same period;** and therefore two circularly polarized rays in opposite directions travelling with different velocities as stated above, are equivalent to a plane polarized ray travelling with mean velocity, and having its plane of polarization rotated at the rate $\frac{\delta}{\lambda v}$ per unit of distance travelled, if δ denote the difference of the two velocities, v the mean of the two, and λ the wave length in the

* Added Oct. 21, 1884.

** Thomson & Tait § 73.

mid. m.* But for lights of different homogeneous colors I found δ to vary as $\frac{1}{\tau}$, that is, as $\frac{1}{\lambda}$. Call it $\frac{A}{\lambda}$. Hence rotation per unit of distance travelled $= \frac{A}{v\lambda}$. And by ordinary dispersion $\frac{1}{v}$ = refractive index $= \mu_0 + \frac{C}{\lambda^2}$. Hence rotation per unit of distance travelled

$$= \frac{A\mu_0}{\lambda^2} + \frac{AC}{\lambda^4}$$

which agrees with the result of observations, showing that $\lambda^2 \times$ amount of rotation per &c., is, to a rough approximation constant and augments as we go up from red to violet.

I have a great improvement however on the mechanical model for gyrostatic effect, which I described in my lecture of Oct. 16. I had thought of it while striving to make something fairly satisfactory of the gyrostatic affair for the lectures but have only succeeded in developing it satisfactorily since their termination. I shall, if possible, write it to-morrow before I sail. I hope at all events to write it during the voyage and post it in time for incorporation in your report.** The same also for metallic reflection &c. and Kerr's magnetic reflection. W. T.]

To-morrow I think we shall see that the anomalous

---

*Remark how very small δ/v is in all known cases, of the Faraday effect in transparent mediums; but how not very small it is found by Kundt (Phil. Mag. Oct. 1884) to be for light passing through an excessively thin film of metallic iron magnetized transversely. This case seems splendidly in accordance with the molecular dynamics of metallic reflection and the transmission of light through metals suggested in my last Lecture, and developed in the addition I am going to send. [See Appendix]

** See Appendix

dispersions and reflections and the heatings that I have been speaking of by the absorption of light passing through a not perfectly transparent medium are all going to be explained simply and well and that this molecular theory has the merit of telling us things we did not know before. It seems not at all improbable that we shall find thin transparent bodies in which the velocity of propagation of light is greater than in the luminiferous ether. If you look at the formulas when it is ready for you, you will see that when T^2 is somewhat less than the shortest of the critical periods κ_1^2, we have μ^2 negative μ^2 is $-\infty$ when T^2 is just less than κ_1^2; decrease T^2 a little more and you get $\mu^2 = 0$; decrease it still more and you get $\mu^2 < 1$, or the velocity of propagation is greater than in the ether. I think we ought to find that phenomenon. I think Quincke found that in some metals ~~the~~ the velocity of propagation is greater than in the ether. There has been very little prismatic examination of the bodies that show anomalous dispersion. It has been alluded to by some of those who have done most in that subject, but there is more to be learned. I think it will be not at all improbable that we shall find zero refractive index and a refractive index less than unity in the neighborhood of some of these critical points. I do not say its a very fundamental phenomenon, but it is worth looking for. Quincke says that there is a very distinct acceleration, showing a greater velocity of propagation in metals than in the luminiferous ether.

What seems to me to be the true theory of absorption is a storing for a moderate time of energy in the attached molecules. Instead of putting in viscous terms in our equations with resistences depending on the velocities, I am disposed to admit no such terms as I have already said and to look for the explanation of absorption in the manner I have indicated. Looking at it in that way, and taking in connection reflection it seems to me that we should have total reflection for those rays whose frequencies are just a little above a critical frequency — rays which are such as to make μ^2 negative. We may put down the mathematics of that for you to-morrow perhaps. That corresponds to a case in which light cannot get into the medium at all and it must be totally reflected unless there is absorption. It seems to me not very improbable that the great proportional amount of light reflected from polished silver surfaces may be explained in that way. Why is so much absorbed and lost in other metals? We cannot tell. But I think that somehow or other, if we take natural suppositions as to attached molecular systems with particles massive enough and lightly enough connected by means of springs, and suitably connected somehow or other by springy connections with the medium, that not only in the neighborhood of critical values, but through a very wide range of frequency of vibration, we shall find a great amount of conversion of the energy into vibrations, i.e. of absorption. There is no real loss of energy, absorption distinctly going to the heating of the body by generation of vibrations in it. I do not despair of seeing an explanation of metallic reflection in this way. I am going to say a little about that to-morrow, but we shall not have a mathematical lecture at all to-morrow.

I want to show you some of this work that

Mr. Morley has gone through. He has found five of the seven roots and the results are most interesting. The roots are 3.4618, 1.0048, .2986, .025561 .007256. I think the 2 roots that are not found are between the two last and the three preceding. It is interesting in connection with the continued fraction, and the form of working I pointed out to you that the U's are, as we know they must be all positive for the smallest root or root of the greatest frequency $\frac{1}{\sigma_2} = 3.4618$. That means that the particles are all moving in opposite directions. For the next root $\frac{1}{\sigma_2} = 1.0048$, the first U is negative, and all the rest are positive. That means that number 1 particle moves in the same direction as number 2 particle, while particles 2, 3, 4, 5, 6, 7 are all moving in opposite directions; and so on.

As to the distributions of energy, taking the successive roots, the particles that have the greatest energy are farther and farther away from the end from which we work. The consideration of the distribution of the energy in these different modes is of vital importance in respect to the application I desire to make in this subject. I thought the working out of an example of that kind would help us greatly, and I am sure we are under obligations to Mr. Morley, for having made these calculations. I hope some of you will not forget another question that I suggested to the arithmetical laboratory because it will throw great light upon the theory of deep sea waves. What I say to-morrow will be upon that subject. I am going to show you that when we attack molecules to the ether, the work done on the medium per period is much less than the energy per wave length and that therefore as front of a succession of waves cannot penetrate into the medium with constant velocity and undiminished amplitude as it does in the familiar case of this wave machine which Prof. Rowland has had constructed for us [Consisting of some 50 or 60 bars attached equi-distantly along a pianoforte wire in the manner already described in the case of the molecular model, and suspended from the ceiling]. There we have waves penetrating with constant velocity and without change of form. Work done by the wave front per period equal to the energy per wave length, is the condition that is necessary and sufficient for the propagation of waves of all lengths at the same velocity, and the same condition is sufficient for the propagation of a pulse without change of form. The question of velocity of groups which was discussed at Montreal, is touched upon here. I do not know whether I can throw any light upon it in connection with Mr. Michelson's observations or not. The thing is of enormous importance in connection with the theory of light, besides being exceedingly interesting in itself as a problem.

Lecture XIX.

We now have (see following page) the problem of the determination of the periods, displacements and energies for the seven particles that I gave you completed. I was under a misapprehension in supposing that there were two roots in a sertain gap. Prof. Franklin noticed that the first root is 3½ times the second, the second, rather more than 3 times the third, the third, about 3½ times the fourth: the fifth is as we now know, about 3½ times the fourth, the sixth about 3½ times the 5th and the seventh about 3½ times the sixth. They are not exactly in that geometrical ratio of 3½ but it is curious that they are not far from being so. I gave you root 3 and 5, and said there were two roots in between. Prof. Franklin said that it was very improbable, and we find that there is another root less than the last root I gave you yesterday.

The maximum displacements in the first mode of vibration corresponding to the greatest value of $\frac{1}{T}$, (that being the frequency in Lord Rayleigh's language) are alternately positive and negative. That must be the case in any system whatever of a similar linear character to this. In the last mode they are essentially all positive. The tendency is to have one fewer change of sign in each successive mode than in the one before it. I cannot give you that as the general rule, because there may be cases in which a node coincides with one of the particles. That is a very common case. In the gravest mode it is obvious that all are swinging in one direction. I will hold this lower particle P at rest, and

Solution for Fundamental Periods, Displacement and Energy Ratios of a System of Spring Connected Particles

$m = 1, 4, 16, 64, 256, 1024, 4096.$

$C = 1, 2, 3, 4, 5, 6, 7, 8.$

By Edward W. Morley, Cleveland, Ohio.

Fundamental Periods Corresponding to Outer Ends of Springs 1 and 8 held fixed

$\frac{1}{T}$ =	3.4618	1.00483	0.29849	0.0880078	0.0255607	0.0072564	0.0014701

Displacement Ratios or values of $\left(\frac{x_i}{x_1}\right)$

x_1	1.	1.	1.	1.	1.	1.	1.
x_2	−.231	1.000	1.351	1.456	1.487	1.496	1.499
x_3	.014	−.341	1.047	1.589	1.761	1.813	1.829
x_4	−.III27	.025	−.431	1.129	1.787	1.997	2.066
x_5	.V15	−.III50	.033	−.511	1.223	1.960	2.216
x_6	−.VIII26	.V30	−.III68	.040	−.581	1.322	2.203
x_7	.XI13	−.VIII51	.V89	−.III81	.045	−.628	1.717

Energy Ratios or values of $\frac{m_i x_i^2}{m_1 x_1^2}$

$m_1x_1^2$	1.	1.	1.	1.	1.	1.	1.
$m_2x_2^2$	.213	3.998	7.30	8.48	8.85	8.96	8.99
$m_3x_3^2$	.II33	1.864	17.54	40.41	49.64	52.58	53.54
$m_4x_4^2$	.V47	.039	11.88	81.66	204.35	255.34	273.14
$m_5x_5^2$	.IX61	.IV65	.28	66.73	382.71	983.10	1157.52
$m_6x_6^2$	.XIV7	.VIII9	.III47	1.62	345.60	1788.13	4968.41
$m_7x_7^2$	.XX7	.XII1	.VII63	.002	8.42	1616.99	12080.04
Sum	1.21	6.90	38.00	199.90	1000.57	4706.10	18542.64

Lecture Notes, October 17th

Comparison of Work with Energies

$$\xi = R \sin \frac{2\pi}{\lambda}(y - Vt)$$

V being the velocity of propogation as modified by embedded molecules

$\frac{\rho}{4\pi^2}$ = density of the ether.

$\frac{e}{4\pi^2}$ = rigidity " " "

I Work done by wave surface in one fixed plane reckoned per unit area of the plane, and per period of the motion

$$= \int_0^\tau dt\, \frac{e}{4\pi^2}\frac{d\xi}{dy}(-\dot{\xi})$$

$$= \frac{R^2}{2}\tau \cdot \frac{e}{\lambda}\cdot\frac{V}{\lambda} = \frac{R^2}{2}\frac{e}{\lambda}$$

$$T = \frac{e}{4\pi^2}\frac{d\xi}{dy}$$

Rate of doing work $= T \times (-\dot{\xi})$

II. Potential Energy of the (distorted) ether, per wave length

$$= \int_0^\lambda \left(\frac{1}{2} T \cdot \frac{d\xi}{dy}\right) dy = \frac{1}{2}\int_0^\lambda dy\, e\left(\frac{d\xi}{dy}\right)^2 \frac{1}{4\pi^2} = \frac{R^2}{2}\cdot\frac{1}{2}\frac{e}{\lambda}$$

III. Kinetic Energy of the (moving) ether per wave length

$$= \int_0^\lambda dy \cdot \frac{1}{2}\frac{\rho}{4\pi^2}\dot{\xi}^2 = \frac{R^2}{2}\cdot\frac{1}{2}\frac{\rho V^2}{\lambda}$$

$$I - (II + III) = \frac{R^2}{2}\left\{\frac{e}{\lambda} - \frac{1}{2}\left(\frac{e}{\lambda} + \frac{\rho V^2}{\lambda}\right)\right\} = \frac{R^2}{2}\cdot\frac{e}{\lambda}\cdot\frac{1}{2}\left(1 - V^2/\frac{e}{\rho}\right)$$

try for the gravest mode. I can almost hit it off — not quite — by merely disturbing the uppermost one; it brings the others with it. That is very nearly the gravest mode. There is a little wiggling to the lowest of the movables, it has not quite got it. It is not quite in order. These two come together at the end of their ranges. They should all be going out together and coming in together. ~~It is not quite in order~~; I will bring it to rest. ~~out together and coming in together.~~ Now we more nearly hit off the gravest mode. But one has a wiggle on it — wiggle is not my word, if you please, I adopt it. "A little wiggle superimposed on the graver mode" tells better the state of the case than more dignified words could. There, then, is the gravest mode with a little wiggle superimposed.

I think some of you will be induced to repeat these experiments, whether Professors or not. See how easily this model is made. Do the work at home with your own hands; then you will have a very interesting piece of miscellany. I wanted to make a fundamental part of our subject (and I only wish we had a few more days to introduce that and some other similar things) the transmission of waves along a row of particles instead of a continuous line. For example, the transmission of a set of waves like waves in a cord along a necklace. Instead of a rope with matter uniformly distributed through it take a necklace with beads strung along it. It is exceedingly easy; we have the equations for it. What we have to do is, in our equations, to take $m_1 = m_2 \ldots$ and $C_1 = C_2 \ldots$, and we get a pretty set of initial conditions, and a charming piece of work that I would have liked to have spent an hour upon, and I think you would have liked it too. Along an infinite row of such particles waves can be propagated in any period longer than the period of that vibration in which every two particles are turning in opposite directions. Start them with equal amplitudes in opposite directions, and think of the time of vibration. You can calculate that from your initial data, one particle with a twisting force C towards a fixed point on the one side and a twisting force 2C towards the fixed point on the other side. The theory leads us to this conclusion that waves in any period less than a critical period cannot be propagated along an infinite row of particles mutually acting each on its predecessor in the series. Equal particles, and equal forces, that is Lagrange's system of linearly connected bodies; and that is a very sweet problem in mathematics — a lovely problem.

If you try to send a wave in a period shorter than the critical period, what is the result? This figures in that paper of mine on the size of atoms. I think some of you may have seen it in Nature. This model of a wave machine is not constructed for illustrating that particular thing because the period is too short. But I give you a hint, if you want a pretty thing. If you are asked to give a popular lecture on waves of light or anything of that kind, you cannot have a better illustration than this to make people understand what you are speaking of. If you make this machine see that the pins have proper obliquities to press the wire in close to the wood. Then cut away the wood where the wire touches it in coming away from the pins. I would suggest that you have the bars very close; so that you can scarcely see anything through them; the vibrations will be prettier. There are three kinds of models one on this plan, then the wiggler, and lastly another one with particles placed at considerable distances apart — perhaps six inches — and with masses, such that the critical period with any particle held at rest may be moderately large.

Then you will see the result of trying to send a set of waves along it smaller than the smallest period for which waves can be transmitted. This wiggler I want you to notice, will not send waves along at all. You get simply a vibration of one particle in the direction opposite to its neighbor. Your finite difference equation for waves has imaginary roots when waves are transmitted, and real roots when the period of the exciter is shorter than the shortest period for which waves can be transmitted. The finite difference equation gives you a simple algebraic equation. Choose the simplest possible quadratic equation – the products of the two roots equal to unity; work that out and you will see that when you have imaginary roots, you have one case, and when you have real roots, the other. A beautiful solution it is. When you have real roots, the displacement of the i'th particle takes the form ρ^i into the displacement of the first particle, ρ being the one of the two roots which is smaller than unity. The general solutions is $C_o\rho^i + C'_o\rho'^i$, $\rho\rho' = 1$. In cases of excitement at one end, and zero motion at an infinite distance, the answer takes, I think, this form, displacement of particle $i = (-)^i C_o\rho^i$, ρ being the least (and positive) root. Turn to the table of displacements and you will see how different that is from the problem of waves along a row of equal particles that I have been telling you about. With equal particles a wave is transmitted through if the period be anything less than the one critical period corresponding to that case. That is not quite the same problem we have here because this is a problem of a finite number of particles, with the remote particle connected by a spring to a fixed body. But we can see how things are going on just as well as if we had an infinite row by taking the first two or three terms. There is in this first mode very little disturbance spread beyond the third particle, indeed it scarcely reaches ~~that~~ that particle. It begins to perceptibly reach the last particle only in the fifth mode.

But the mass of this last particle is so great in comparison with that of the first that its energy is 8 times the first, with only $\frac{1}{20}$ of the displacement.

I have written down some things I wish particularly to speak to you about, I will tell them to you beforehand. First, the pressural wave, sometimes called the surface wave. It is a surface wave only when it is a condensational wave. The subject I want to speak upon is, the condensational wave, the annulment of it into a mere pressural wave and the nature of this pressural wave. We have spoken of it; you have seen it in the formula; I do not know that you have all got it clearly into your heads what it is. Next, I want to show you, very roughly, the formula of reflection and refraction with vibrations perpendicular to the plane of the three rays. Third, I want to speak about the aeolotropy of inertia first suggested by Rankine, afterwards, independently, by Lord Rayleigh, to account for differences of velocity in different directions, manifested by double refracting crystals. Fourth, Stokes negative to that interesting hypothesis. Fifth, just a very brief summing up of the points of difficulty, and then we will go to our little sheet of lecture notes.

I can best explain the condensational wave by showing you something about waves in general in <u>an elastic solid</u>. Let the space below this line

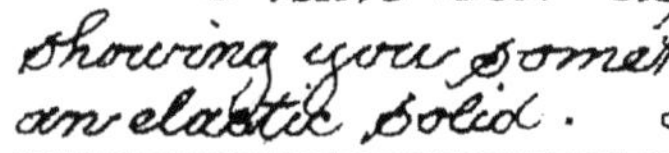

be a portion of an elastic solid. We will think of the propagation of waves into it or vibrations either; the mathematical solution is very indifferent to vibrations or waves. The general formula of yesterday is just as ready to be converted into a formula for vibrations as it is into a formula for waves. But I want, without the formula at all, to think of the propagation of a wave in such a medium. Suppose in the first place, you disturb the surface and hold it disturbed. There will be a certain static problem to solve for the disturbance of the surface. I rub away the original boundary and leave this wave boundary. The

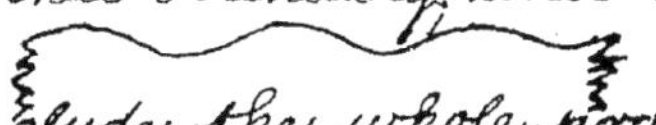

problem of the waves would include the whole problam, and it is very easily worked out. Make ever so slow or sudden static disturbance of any given shape: that problem is not very difficult to work, and is very interestong as a problem of dynamics with a view to physical applications, and it is also valuable on its own account.

Now apply a corrugated rigid form to the surface, and slip that form along at a certain speed. If we slip it along too slowly no waves will be sent into the solid at all. The velocity of propagation of waves is, you know $\sqrt{\frac{n}{\rho}}$ or $\sqrt{\frac{(k+\frac{4}{3}n)}{\rho}}$ according as we have distortional or condensational waves, irrespective of the wave length, — because we are not in molecular dynamics just now, we are in molar dynamics. If we slip our form along at a less speed than the velocity of propagation of a wave, no waves at all will be sent into the medium.

There is a certain charm about the mathematical analysis that gives us the general solution of a problem like this by considerations of mixed real and imaginary quantities. If you calculate the effects of applying a form and moving it at a speed less than the velocity of propagation of a wave in the medium, you will have the result in the form of exponentials with real indices — quite analagous to our problem of a finite number of particles, where we have real roots of the equation with nothing spreading into the interior. Here we have different cases. The most difficult case is for vibrations in the plane of the board. The second case is simpler.

I, repeat again, get the effect of a sudden shock upon the medium — of course, if you twist it out of shape, you send an earthquake through it, but without twisting it out of shape, give it a shock, or whatever you may call it, just as I do when I displace this handle up of our model and cause it suddenly to begin performing a simple harmonic motion. Then get the simple harmonic motion again which every particle performs consistently with the surface being affected by a corrugated rigid form carried along at a constant rate. Our formulas of yesterday are only adapted to giving us the simple harmonic motion; and that is what we are considering – the simple problem of real periodic waves. There can be no periodic waves sent into the medium if the speed of the form is less than the velocity of propagation of the wave — no distortional wave if the speed be less than $\sqrt{\frac{n}{\rho}}$, and no condensational wave if the speed be less than $\sqrt{\frac{k+\frac{4}{3}n}{\rho}}$. We might work the thing out, and a very pretty problem it is.

*[Referring to the motion of a corrugated rigid form along the surface of an elastic solid, I said "greater than" when I should have said less than.† The true statement is as follows:

I (Applicable to vibrations either in the plane perpendicular to the bounding plane and containing the direction of motion of the form — the plane of the board — or perpendicular to that plane).

If the velocity (V) with which the form is carried along is less than the velocity (v) of a distortional wave, no wave will be propagated inwards: only a disturbance of which the magnitude diminishes from

* Added Oct. 21, 1884.

† This has been corrected in the report.

the surface inwards according to the logarithmic or exponential law.

II (Vibrations in the plane of the board)

If the velocity (V) of the form exceeds that of the distortional wave (v) but is less than that of the condensational wave (u) a distortional plane wave is propagated inwards, but no condensational wave. The inclination of the wave front of the distortional wave to the bounding surface of the medium is $\sin^{-1}\frac{v}{V}$.*

III (Vibrations in the plane of the board.)

If $V > u$, two plane waves are propagated inwards – distortional and condensational – the inclinations of their wave fronts to the bounding surface of the medium being respectively $\sin^{-1}\frac{v}{V}$ and $\sin^{-1}\frac{u}{V}$.

This subject ought to be carefully and thoroughly illustrated by diagrams, showing the wave fronts, and the corrugated lines of particles which are in straight lines when undisturbed – all this for vibrations both in and perpendicular to the plane of the board. W.T.]

The problem of reflection and refraction is a small part of this matter. It is more interesting as a problem of mathematical dynamics than anything else. I was saying to Stokes that I wanted this worked out more than it had been done before. He said, "Cui bono" I say there is the cui bono: it is interesting and instructive to work it out. We are all forced to feel that we are rather in a hole – I will not call it the slough of despond because we do not despond, really, as to the explanation of refraction and reflection; and although this will not explain refraction and reflection, let us see what it will do. Books on dynamics could well be devoted to work of this kind. If I am able to go on with the work on Natural Philosophy

---

* Omit the restriction to $< u$, and II becomes unqualifiedly applicable to vibrations perpendicular to the plane of the board.

that I have in mind, I intend to make this investigation.

The question of applying a form and moving it along, and so on, does not exhaust the data for this problem. Our conditions are, a certain form slipped along with certain geometrical conditions to be fulfilled as to the change of shape of the surface, it being always made to fit the form. That will correspond to our first set of equations $\xi = \xi'$ developed yesterday. But with respect to the horizontal component, the form may drag the particles with it. You may vary your data thus: let there be a stated tangential force between the form and the solid at every point. Let the form be so constituted that while it is being moved along it will shove back in some places and shove forward in other places, producing a given distribution of tangential force all over its surface. The given distribution of tangential force must vary according to a simple harmonic motion in order that we may get a simple problem. It must vary as the sine of the angle corresponding to the variation, or it must be expressed as an exponential logarithm.

We need not go further with that sort of problem. You can see what it is. Without thinking of it as a corrugated form applied to the medium, think of it that you act upon every element of the surface of a medium with a normal and a tangential force after you have given it any displacement you please constituting a <u>given</u> set of waves in the medium. We took that as a reason yesterday for making a coefficient unity. Somebody might have said, "Why do you not take the incident ray as given, and the refracted and reflected rays as the unknowns?" I answer, in the lower medium there is only one plane wave, unless there be a condensational wave. It is convenient then to take that medium in the first place and the other in the second place; and furthermore, we get a perfect symmetry if we take unity say for the coefficient of the refracted wave and then leave two quantities for the reflected wave. The two ratios of the three things is all we want.

Have you ever thought (it is a curious enough explanation this) what sort of an arrangement would have to be made in order to have one incident ray giving rise to a refracted ray, and a quasi-reflected ray? Think of the case thus: reverse the motion of every particle in the problem as put here. We cannot produce such a thing, but there is not the slightest difficulty in imagining it. If the motion of every particle concerned was to be reversed, the refracted wave would travel back, our originally incident ray would travel back, and the reflected ray would travel in the corresponding reversed direction. That is a sample of what is introduced in the mathematical treatment of all such questions; but as to getting a source of light with its vibrations and relations so timed that that would be the result — there is no such thing. You will notice, also, that the work done by the wave front in any part of the incident wave per period is equal to the energy per wave length in the first medium, and according to our formula as worked out, that would hold for the second medium; so that the sum of the energies per wave length in the reflected and refracted rays is equal to the work done per period in the incident ray. In reversing, we must take that into account, so that we must supply a state of energy at the surface in order to make things come out in the way I have stated.

I will now put in a medium above our medium which we have been considering. The displacements in the interface of the two mediums are the same — not merely the normal components of the displacements, but the tangential components. That gives two particular equations. The upper medium pulls upon the lower with normal and tangential components of force. You might imagine other cases. Although it would be not at all an interesting problem, you might say, let there be a possibility of finite slip between one and the other, or rather you might imagine the two not ~~to~~ cohering together but to be separated and to be perfectly smooth. The result would be zero tangential force in each medium, giving two equations, with a third and a fourth equation, viz., normal components of displacements and normal pressures in the two mediums equal. It is not an interesting view, because a finite slip between the two mediums is an inconceivable arrangement for our optical application at all events. Yet I do not think it is a less interesting problem merely as a problem of mathematical dynamics to suppose the two mediums to be separate and perfectly smooth. We cannot do away with the equality of normal pressures—we cannot get a mathematical problem according to that—because it would be inconsistent with harmonic motion. It is not inconsistent with reality, as anybody who has tried to ring cracked glass will see. Stroke cracked glass and notice the jarring. That comes from the cracks bending and slipping together. These are not the kind of problems I want to look at now.

You see that the problem we are solving comes out wonderfully simple when put into the form of a problem with only four unknown quantities by means of imaginaries. I put down yesterday what our condensational wave becomes when realized. $\varphi = \epsilon^{-bx}\{C\cos(by+\omega t) + D\sin(by+\omega t)\}$. I want to get quit of this. I do not want to go into details, but will just call your attention to the last equation in yesterday's paper, and the form I put it in afterwards.

that for all angles of incidence between zero and $\sin^{-1}\sqrt{\frac{n}{k+\frac{1}{3}n}}$ we have a plane condensational wave going into the interior with the distortional wave, and also a reflected condensational wave. The only way to get quit of that for all angles of incidence is to suppose $k = \infty$. If k be very large in comparison with n a condensational wave will only be generated between zero and a very small angle, viz: $\sin^{-1}\sqrt{\frac{n}{k}}$. I do not know as to our right to say that k is infinite, although there is no doubt we have a right to say that it is very large. Stokes went into that very fully in his report on double refraction and has given really the substance of every conceivable illustration of it. He shows that in every reflection and refraction, at all events with not too great obliquities, there will be a condensational wave generated from light falling on a body which consists of merely distortional waves; and he shows that according to the supposition that Cauchy made, which assumes Poisson's and Navier's ratio for elastic solids, that the condensational wave has a magnitude of very considerable energy compared with the distortional wave. Even if the ratio of n to k were enormously less than Poisson's ratio would make it, the energy of the condensational wave would still be so much as to produce immense effects. If you take an exceedingly intense light, that would produce a condensational wave of small energy in comparison to its own, perhaps a ten-thousandth of its own energy. But take sunlight falling upon a piece of glass — waves having a ten-thousandth of the energy of sunlight would have still very large energy compared with ordinary light and that again, going at a velocity different from what we know — enormously greater — would, in falling upon a body, develop distortional light; and we should have distortional light, ~~and we should have distortional light~~ springing up in places where there was no visible cause for it. We know of no such phenomenon. We are perfectly certain that if there is any such phenomenon it is of exceedingly small energy compared with light. I think we may safely say, whatever condensational wave there may be, its energy cannot amount to more than one-hundred-thousandth of the actual energy of the distortional light that produces it. It might or might not amount to a much larger proportion than that. But all I say, you understand, is that we have no such agency going about through the universe — enormous quantities of it coming from the sun with sunlight — from the fact that we have no trace of it in nature, and no evidence of such a force coming from the sun. There being no trace of it resulting from the combination of materials in practical experiments, we infer with certainty that if there is a condensational wave at all, it is of excessively small energy in comparison with the energy of the distortional wave accompanying it or giving rise to it.

Therefore we say k is practically infinite, and our attempt to introduce a condensational wave has been a woful failure. Make now $\mathfrak{b} = b$ and the result is $\phi = C\varepsilon^{-bx}\cos(by + \omega t)$, which is simply the well known expression for the displacement potential of deep sea waves corresponding to wave length $\frac{2\pi}{b} = b$. If you look at our formula of yesterday, you will see that $\frac{2\pi}{b}$ is the length from crest to crest. In our expressions you will see that the coefficient of y is the same throughout. In each particular expression it is $= 2\pi \sin i \div$ by the wave length, which corresponds to the wave length l in the extreme case of grazing incidence.

Take the extreme case of grazing incidence, and if the wave length in the refracting medium is longer than in the other we have a wave travelling in one and in the other not, and it comes out a case of total internal reflection. If the wave length is shorter in the lower medium, the true state of the case will be this: We would have a vertical set of wave fronts in the upper medium and a case of light refracted into the lower medium with inclined wave fronts, the wave length being shorter: We have $\sin i = \frac{1}{n} \sin i'$, and i being $90°$, we have $\sin i' = \frac{1}{n}$; the well known case. As to the ~~total internal~~ treatment of total internal reflection that comes out with extraordinary ease from the analytical method, as you all know.

The condensational wave has become no longer such by the supposition $\bar{b} = b$; it is what may be called a pressural wave. Lord Rayleigh calls it a surface wave. It is a wave that spreads into one medium and the other so as to produce disturbances in condensations through a range comparable with the wave length — comparable with this quantity $\frac{1}{b}$, which is comparable with the wave length for any angle of incidence of considerable obliquity. I have put down the form for the upper medium. For the lower medium it is $\Phi = C'\varepsilon^{bx} \cos(by + \omega t)$. x is negative in the lower medium, which justifies the change from $+b$ to $-b$. i.e. ε^{-bx}, which occurs in Φ, occurs also in its differential coefficients, and is therefore the coefficient of diminution of the displacements as we recede from the interface. Take $x = \pm l = \pm\frac{2\pi}{b}$ and we have a coefficient $\varepsilon^{-2\pi}$ — what is the magnitude of that? In my own classes when I am lecturing on this subject, I ask my boys to write in the first page of their note books the values of $\varepsilon, \varepsilon^2, ..$ $\varepsilon^{\frac{1}{2}}, \varepsilon^{\frac{1}{4}}, ..$ also $\varepsilon^{\pi}, \varepsilon^{2\pi}, .. \varepsilon^{\frac{1}{2}\pi}, \varepsilon^{\frac{1}{4}\pi} ...$ Thackeray says, no person ever calculates his own logarithms. Quite wrong; every mathematician calculates his own logarithms; he must calculate them in order to have them. Thackeray did not know that. But notwithstanding it is not true, that expression is a good one for illustrating the subject. I only remember two figures of the value of ε. It is about 2.7 – that raised to the power 2π is a large number. $\varepsilon^{-2\pi}$ then is a small fraction. Our displacements are then very small when $x = l$. Take $x = 2l$ and the coefficient of diminution is excessively small.

This is precisely the case of a deep sea wave, and you see that the motion of the water at a depth of the wave length is very small. Even at half the wave length the coefficient is $\varepsilon^{-\pi}$; or at a depth of half a wave length the disturbance is only about $\frac{1}{27}$ of what it is at the surface. The diminution is enormously rapid. That is exactly the case with this pressural wave. It produces a disturbance in each medium which is sensible at distances comparable with the wave length; insensible at distances a considerable multiple of the wave length. There is no difficulty in thinking of pressural waves in an incompressible solid, an elastic jelly for instance. We cannot have a pressural wave at all in the interior of an infinite incompressible solid. It must get away somewhere. If it is free on one side, there is no difficulty about it. I must withdraw that remark that a pressural wave cannot originate in the interior of an incompressible solid. Move a body about in the interior of such a solid, and you have a solution – Φ for the case of k infinite is a solution with definite displacements corresponding to a pressural wave. But none of that kind of effect appears at distances from the source considerable in comparison with the wave length. We can not prevent the introduction of this pressural wave, or quasi-water wave as others call it, in order to allow of the two components of displacement and two components of force on the two sides of the interface being equal. The extinctional formula by which Cauchy gets rid of the condensational wave, and those hypotheses of Neumann and MacCullagh that are still (as if there was any importance or weight to be attached to them) spoken of as if they were theories, are merely mistakes. I am a little aroused because I read, not two hours ago, an article in the Compte Rendue, by a new name, taking up with all gravity Neumann's theory and MacCullagh's theory and giving great weight and importance to them, finding that they come within an exceedingly small fraction of so and so, and so on. Go into it as analyzed by Lord Rayleigh, and you will see that their theories consist in introducing conditions that are inconsistent with two portions of matter pressing against one another with equal force, one pressing against the other with the same force that the other presses against it. We ask nothing more than that action and reaction are equal and opposite at the separating surface together with continuity of matter. Those are the only principles, notwithstanding the four principles of MacCullagh that I read to you the other day from Lord Rayleigh. These are the only principles, that action and reaction are equal and opposite, giving two equations, and that matter is continuous and does not slip, giving two more. These are comprised in one, viz: mutual impenetrability of the two homogeneous mediums. I think we have spoken of that bête noir sufficiently. Leave it alone and you see it is a good enough animal after all.

I wanted to give you the reflected and refracted

rays; but I need not do so because most of you have access to Lord Rayleigh's paper on the Reflection of Light from Transparent Matter. You see how charmingly short it is: there is the whole of it. Read that and then look at the conclusion. Green makes his simplification $n = n'$ entirely too soon; otherwise he might have got this result and said "This is what I got in the case of reflection of sound." Nothing could be simpler than this. $n = n'$ is a very slight simplification for the comparatively not very difficult case of only four unknown quantities. In case you do not have Lord Rayleigh's book at hand note this if you think it worth while: Ratio of amplitudes of reflected to incident vibrations = $\left(\frac{\tan i'}{\tan i} - \frac{n'}{n}\right) \Big/ \left(\frac{\tan i'}{\tan i} + \frac{n'}{n}\right)$, which becomes Green's sine formula for $n = n'$. That lovely formula, as I call it, is given first so far as I know by Lord Rayleigh. I am pretty certain that he is the first who has given it correctly, because I know of no other writer except Green who worked at this problem without introducing impossibilities that vitiate the whole affair; and Green did not do it. Vibrations, then, perpendicular to the plane of incidence for two elastic solid media—no matter whether compressible or incompressible—give the same law as to intensity of reflection as two fluids destitute of rigidity, (and therefore giving us a case in which the vibrations are essentially in the plane of the three rays). Vibrations purely compressional in a medium without rigidity are essentially in the plane of the three rays and give identically the same expression for the ratio between incident and reflected vibrations as does an elastic solid with vibrations perpendicular to the plane of incidence.

Having obtained this formula, Lord Rayleigh

takes up the cases. About four days ago, I got hold of the thing wrong side up and it was only a few hours ago that I took up the cases right, and I find everything is true, interesting, intelligible and instructive.

Case I is Green's $n = n'$, which gives his sine formula $\left(\frac{\sin(i'-i)}{\sin(i'+i)}\right)$ for the ratio of reflected to incident ray. Case II is MacCullagh's, $\rho = \rho'$. MacCullagh is a very clever and able man, but he ignored dynamics vitally in the most peculiar parts of his work. We have $\frac{n'}{\rho'} / \frac{n}{\rho} = \mu^2$; $\frac{n}{\rho}$, $\frac{n'}{\rho'}$ being the square of the velocities in the two mediums and μ^2 the refractive index. Take then $\rho = \rho'$ and we have $\frac{n'}{n} = \frac{1}{\mu^2} = \frac{\sin^2 i'}{\sin^2 i}$. Substitute that in the tangent formula, and it reduces to $\frac{\tan(i'-i)}{\tan(i'+i)}$. We have therefore this case of equal densities and unequal rigidities giving complete extinction at the angle of polarization. I have told you about that. It is satisfactory as a mathematical problem, but it is a failure for what we wish to account for in the theory of light. But we must stop here I am afraid.

Lecture XX.

I have down next in my notes Rankine's very beautiful suggestion of aeolotropy of inertia. We want to explain aeolotropy in a crystal. We know that the velocity of propagation depends on the direction of vibration and not on the plane of distortion. Rankine's idea was this: let there be connected with the ether, or imbedded in it, gross molecules. I do not say ponderable or imponderable, but I use the word gross not meaning to throw any obloquy on them but simply to say that they are large. I do not say that I am giving Rankine's way of doing it. He mixes it up with molecular vortices and so on, and it is the kind of molecular vortices that we can not very well get an idea of. I do not think I would like to suggest that Rankine's molecular hypothesis is of very great importance. The title is of more importance than anything else in the work. Rankine was that kind of genius that his names were of enormous suggestiveness, but we can not say that always of the substance. We cannot find a foundation for a great deal of his mathematical writings, and there is no explanation of his kind of matter. I never satisfy myself until I can make a mechanical model of a thing. If I can make a mechanical model I can understand it. As long as I cannot make a mechanical model all the way through I cannot understand, and that is why I cannot get the electro-magnetic theory. I firmly believe in an electro-magnetic theory of light, and that when we understand electricity and magnetism and light, we shall see them all together as part of a whole. But I want to understand light as well as I can without introducing things that we understand even less of. That is why I take plain dynamics. I can get a model in plain dynamics, I cannot in electro-magnetics. But as soon as we have rotators to take the part of magnets, and something imponderable to take the part of magnetism and realize by experiment Maxwell's beautiful ideas of electro displacements and so on, then we shall see electricity, magnetism and light closely united and grounded in the same system.

Suppose here a massless rigid lining of our ideal cavity in the luminiferous ether. Let there be a massive heavy molecule inside, with fluid around it. The main thing is, that this molecule, which only affects the effective inertia of the ether by adding its own mass to the moving mass of the ether, has aeolotropy of inertia. Imagine this spherule moving first in a horizontal direction. The effective inertia of this sheath will be altered if it moves to and fro in a vertical direction, there being by hypothesis liquid between it and the ether. The density of this mass must be greater than the density of the liquid, that is all. If there is danger of its coming to the sides of the cavity let there be springs to keep it in place if you like but let its connection with the lining of the cavity be in the main through fluid pressure. Then its effective inertia is different in different directions. This fluid lining seemed to hit off the very thing we wanted. Now comes Rankine's want of strength. He cut around the edges of it, and I think, rather jumped at it, and put down a wave surface the same as Fresnel's and said that it came to that. But alas, Stokes (long before Lord Rayleigh suggested it) showed that it would give a different surface from Fresnel's. Lord Rayleigh, in repeating Rankine's suggestion, showed his strength where Rankine was not so strong, in mathematical powers of grappling with a different dynamical problem. Lord Rayleigh is a man who grapples with a difficulty and sees how much he can do with it. He puts it aside if he cannot solve it; but he never shirks it. Rankine was not a mathematician in that sense at all. Lord Rayleigh finds, not Fresnel's wave surface, but a wave surface differing from Fresnel's by certain terms appearing in reciprocals instead of directly. Lord Rayleigh could not pick up a thing of that kind without seeing the end of it, and he says in conclusion:*

"Between the theory here advanced and that of Fresnel observation ought to decide; but it does not appear that any experiments hitherto made are competent to do so. As Prof. Stokes points out, all the measurements which are to be combined in one calculation should refer to the same specimen of the crystal; otherwise an element of uncertainty is introduced sufficient to render the application of the test ambiguous. Should the verdict go against the view of the present paper, it is hard to see how any consistent theory is possible, which shall embrace at once the laws of scattering, regular reflection, and double refraction."

In the course of that paper Lord Rayleigh finds that Stokes had written that up and he is greatly

---

*Philosophical Magazine, June (Supplement) 1871, "On Double Refraction".

surprised. The way he refers to Stokes is rather interesting: "I had got about as far as this in my original work when, on reference to Prof. Stokes's report, I was greatly surprised to find allusions to a theory of double refraction mathematically, if not physically, identical with that here advanced." After insisting on the importance of precise measurements, he says:– I will not read all that. Here is something: "Were the law [says Stokes] of wave velocity expressed for example by the construction already mentioned having reference to ellipsoid (12), the wave surface (in this case a surface of the 16th degree) would still have plane curves of contact with the tangent plane, which in this case also, as in the wave surface of Fresnel, are, as I find, circles, though that they should be circles could not have been foreseen." That is in respect to conical refraction, which Stokes says is thus no test of Fresnel's construction. Stokes told me of all this. It was he who first called my attention to the fact that Rankine was doubtful. He had not made his experiments then; but sometime after he told me of them. It seemed to me that they were experiments of very great accuracy, and I implored him to publish them. It was very hard to get him to do it. Every time I went to Cambridge I asked him to publish his results. Finally he did, and here is the whole of it, just 12 lines in the Proceedings of the Royal Society, June, 1872, under the title "Law of Extraordinary Refraction in Iceland Spar," and he has never published a word more about it. "It is now some years since I carried out in the case of iceland spar the method of examination of the law of refraction which I described in my report on Double Refraction, published in the Report of the British Association for the year 1862, page 272. A prism approximately right angled isosceles was cut in such a direction as to admit of scrutiny across the two acute angles in directions of the wave normal within the crystal comprising respectively inclinations of 90° and 45°, to the axis. The directions of the cut faces were referred by reflection to the cleavage planes and thereby to the axis. The light observed was the bright D of a soda-flame.

"The result obtained was that Huygens construction gives the true law of double refraction within the limits of errors of observation. The error, if any, could hardly exceed a unit in the fourth place of decimals of the index, or reciprocal of the wave velocity, the velocity in air being taken as unity. This result is sufficient <u>absolutely to disprove</u> the law resulting from the theory which makes double refraction depend on difference of inertia in different directions.

"I intend to present to the Royal Society a full account of the observations; but in the meantime, the publication of this preliminary notice and the result obtained may be useful to those engaged in the theory of double refraction."

That was in 1872. 12 years have passed and nothing more has been published. You should be grateful to me for getting so much; you owe it to me.

I have next to consider some of the difficulties. What are they? Without the question of double refraction at all, consider simply the problem of reflection and refraction at the separating surface of transparent mediums. Take the theory that you know work out every detail on the supposition $n=n'$; and that gives us at best only a rough approximation to Fresnel's results. They do not come near expressing the extinction at the polarizing angle.

Of one thing we are sure, the only way of coming

at all within one-hundred miles of explaining the known facts of polarization is by supposing the vibrations to be perpendicular to the plane of the three rays. We are certain that if light is to be explained by the problem of an elastic solid, that the vibrations must be perpendicular to the plane of the three rays unless we are to alter our facts altogether. I tried it with my molecules, and it makes no difference. My molecules give exactly the same result as the theory before you—no modification whatever. We cannot help ourselves at all by the molecules.

Then comes the difficulty (if you call it that) of making the line of vibration perpendicular to the plane of the three rays in the case in which we have no approach to extinction of the reflected ray. The difficulty is to get so near an approach to extinction as we have at the polarizing angle for light vibrating in the plane of incidence, and to explain the results of observation or the supposed results of observation that we have on the subject. These, according to Jamin's experiments, are very curious and noteworthy. According to his experiments there is a certain critical case for refraction, in which the refractive index is 1.4. If they are all right there would be perfect polarization for refractive index $\mu = 1.4$ and the phase going opposite ways from that—I am speaking very badly, but you will understand; the order of things as regards change of phase would be opposite for refractive index exceeding 1.4 to what it is for refractive index less than 1.4. Something like that results from Jamin's work; but his work was done a long time ago, and some people think it not altogether trustworthy. I do not know as Jamin himself would be fully satisfied with it now.

More work is wanted in the subject. Do not

let us break our wings in battling against, and in trying to explain, facts which may not turn out to be facts. We can work on the theory, and try to get all we can out of it with its 21 coefficients; but let us also work together and get some of the facts. I hope you will all make observations on the polarization of light. That expression <u>elliptic polarization</u> should always be coupled with elliptic polarization in reflected light when the incident light has been plane polarized with its plane neither in nor perpendicular to the three rays. Elliptical polarization is a confusing expression — find what is understood by that. Somebody must do it; I hope some of you will do it. Make also photometric experiment as to the quantity of light. Prof. Rood has made some splendid experiments of that kind. I meant to speak of those yesterday instead of my own rude experiments. He found for reflection of light from one or two substances at direct incidence, a fulfilment of Fresnel's formula $\left(\frac{\mu-1}{\mu+1}\right)^2$ to within a fraction of a per cent. He made experiments on several bodies but has not published them except for ground glass. — Do make him publish them for iceland spar and plane glass. Anything from Rood is certain not to be rude. Like Stokes, he was not satisfied and did not publish his experiments although he made them ten or twelve years ago. After what I have obtained from Stokes, I hope all of you will try and extract the results from anybody who has good things in the shape of results.

I made many years ago a measurement of the celebrated v, the number of electro-static units in an electro-magnetic unit. I have just heard that the measurement has been made here with the whole system of apparatus and with the accuracy applied to electro-static measurement, which seems inconceivably superior to any measurements that have been made anywhere else so far as I know. I intend to get it for the Royal Society of London, which will not preclude its being published in any American publication.

The explanation of polarization at all by reflection, that is a difficulty. After that comes the other difficulty to explain double refraction; to find out how we can get it reasonably without introducing a fallacy of any kind, without introducing some other feature that is contrary to observation: to account for difference of velocity in different directions in a crystal by such a dynamical theory that the velocity of propagation shall be a function of the direction of vibration, and not of the direction of the strain. To read Rankine's splendid failure in this is most instructive and valuable

If you were to ask me what other difficulties there were in the undulatory theory of light, I would say, I do not know that there is any other difficulty. The only other one is the old difficulty of the ether — how the planets can go through it, or, how the molecules of the kinetic theory of gases, going at velocities of from one-hundred to five-hundred meters per second (say half a kilometer per second) can go through it without any resistance, so far as we know, and that yet the maximum velocity of the molecular vibrations which produce light must be a small fraction of 300,000 kilometers per second, the velocity of light. The velocity of the vibrating molecules might amount to $\frac{1}{50}$th of the velocity of light; more probably it is not a thousandth of it. Probably in faint light it is not a three-hundred thousandth of it, or not more than kilometer per second. You see I am taking you into my confidence. I am concealing nothing from you that I see. Here we have the particles going with a velocity of half or a quarter of a kilometer per second in the kinetic theory of gases, and yet we have the molecules creating waves of light by vibrations of a velocity ~~veloc-ity~~ which may not be more than one kilometer per second and cannot probably be as much as a thousand kilometers per second.

I only put the thing before you. Many of you have been thinking of this, no doubt, as a difficulty. I do not want to gloss over anything. But putting this aside, let us come down to ordinary matter. If you make a vibration in glycerine quick enough it will act like a perfectly elastic solid. I do not speak of the velocity of the vibration, I mean the period of the vibration. If the period of the vibration is short enough, I suppose glycerine would act like a perfectly elastic solid. Again, Maxwell's kinetic theory of gases leads us almost to say that for quick enough motions of a molecule in a crowd of molecules—motions by which the theory is explained—we may have a quasi-elasticity as of a solid coexisting with the gas.

But I fall back on glycerine. I tried last winter a new kind of a galvanometer, and I made a woful failure of it, I am sorry to say. I made very many uses of glycerine in checking the vibrations of the needle. The needle would, however, attain its full velocity, make two or three oscillations about a false pole and gradually come back. I could not look at that without being taught by it that the difficulty of a luminiferous ether would turn out not to be a difficulty at all. It is the shortness of the period of the to and fro motion in the luminiferous ether that allows it to act as a perfectly elastic solid for the luminiferous vibrations. For motions of particles of corresponding space not much greater, or perhaps of equal or less space, Here is a perfect line with respect to absolute velocity when the force applied to a molecule acts for a long enough time to get it into motion. Why does a collision between molecules in the kinetic theory of gases give rise to velocities of one or two kilometers per second, or change the velocity one or two kilometers per second? Answer, because the whole time of collision is enormously greater than the four hundred million millionth of a second or than the slowest of the vibrations that Langley has found. In a paper that I have from Langley—I want to speak of it, it is so interesting—he has stated that as 17 times the period of sodium light. Make it 20 times: that gives the rate of 20 million million vibrations per second as the most sluggish vibration we know of in light and radiant heat.

The medium's being perfectly elastic for the to and fro recover nces of motions in the 20 million millionth of a second is perfectly consistent, it seems to me, with its being like a perfect fluid in respect to forces acting perhaps for one millionth of a second.

Imagine what is the force of the collision between molecules. Take two billiard balls, and not allowing for the heat of collision, we can calculate the force of it roughly from our knowledge of the elasticity of the materials. Now, imagine the molecules of oxygen and nitrogen to be about as hard as billiard balls. I think if we only were to see the thing as it is, the collision between molecules on the kinetic theory of gases would appear very gentle influences. Two molecules would come slowly together and be gradually stopped; and if you were to think of the viscosity in relation to all this and calculate it out, you would see ~~the~~ relations we cannot stop to take up just now. But compare that with a to and fro motion twenty million times as rapid. A million is not inconceivable; but it is a tremendous number. Think of once per second as compared with 30 times per second, and you need not think it incredible that the medium acts as if it were perfectly elastic relatively to one vibration and perfectly yielding with reference to the other.

Our molecular theory will fit this. Go back to our spherical molecule with its central spherical shells — that is the rude mechanical illustration, remember. I think it is very far from the actual mechanism of the thing, but it will give us a mechanical model. By working at it, and helping ourselves by such work as this of Prof. Morley's we shall see how every sequence of waves leaves a little more and a little more of energy in the gravest modes of the compound molecule until the energy is absorbed in modes of which the period is perhaps the millionth of a second instead of the 20 million millionth, or the 400 million millionth of a second. Think of the molecules, while they are doing work for light, as also moving about with a velocity of as much as a kilometer per second, say. Well, two of them come into collisional distance and one gives the other a gentle shove in the course of a millionth of a second and causes it to change its speed. Part of the energy that these molecules had from light vibrating at the rate of 20 or 400 million million times per second has been got into the form of long vibrations — so long that when the two come into collision they give to one another the gentle kind of shove required for the kinetic theory of gases.

Thus we can see perfectly how absorption will lead us down through fluorescence, phosphorescence, the heating up of the molecules so that they will give it out again by radiation all around through the ether, and then again still lower degradations, down to the sluggish vibration according to which, two molecules, swinging something like this ⟷ their centres going one way and their shells the other, come together in the period of a millionth of a second, gently shove one another, and go off in other directions, adding their inertia to the velocity or taking it from the velocity or turning the course around at right angles. Thus I can see how our compound molecules act not only to increase the temperature when you increase the pressure according to the kinetic theory, but how the same molecules act to give us fluorescence and phosphorescence and then again the radiant heat from a body which is heated by rays passing through it.

I intended (but the time is too short to carry out that intention) to have worked out a mechanical model for sodium light. I will tell you how to do it so as to show quite an exceedingly sharp effect — as sharp as the two D lines are shown in Prof. Rowland's spectrum. If we had a day or two longer, we would hang on our particle M, a little pendulum — we would have to invoke gravity to help us here. If we are too proud to use gravity we can hang on a little springy molecule whose vibration is a certain period. Stick on beside it another springy molecule whose period varies by $\frac{1}{800}$th or a thousandth from the first and another whose period is ever so little compared with either — Say one whose period is $\frac{1}{899}$th of a second and another whose period is one second exactly. Let these be so small that they produce no sensible effect until the period of the vibrator is within an hundred thousandth of either. Then it will begin to be enlivened up, and begin to make vibrations that will tell. While it is within an hundred-thousandth of the period of one, the period of vibration differs a hundred times as much from the period of the other and the energy of the vibrations produced in the other will be enormously smaller. Think then of adding to our first particle two molecules, with the period of one within a thousandth part of the period of the other, and another whose period is ever so little and in saying good bye to this illustration we will have a perfect model of a molecule that will produce sodium light, and produce the effect that is produced by sodium vapor upon light.

I have brought a book which I intended to make a subject of our lecture. I am afraid it will be passed over. The book is Stokes' paper "On the Metallic Reflexion exhibited by certain Non-metallic Substances.*" I only wanted to tell you that this molecular theory explains the colors of aniline and this wonderful thing that Stokes experimented on — this safflower-red. I wanted to read about the bright lines in the light reflected from safflower-red discovered by Stokes. I was thinking about this three days ago, and said to myself, there must be bright lines of reflection from bodies in which we have these molecules that can produce ~~all~~ intense absorption. Speaking about it to Lord Rayleigh at breakfast he informed me of this paper of Stokes and I looked and saw that what I had thought of was there. It was known perfectly well, but the molecule first discovered it to me. I am exceedingly interested about these things, since I am only beginning to find out what everybody else knew, such as anomalous dispersion and those quasi-colors and so on. There is no difficulty about explaining these things; we can predict them from the consideration of the molecule without

* Phil. Mag. Dec. 1853.

experimental knowledge. And here again is a thing that suggest itself to me, that most probably there are bodies in which light is propagated faster than in the luminiferous ether.

I wish we could go into the dynamics of that but we cannot. Take our old formula that we had about a week ago, μ^2 = so and so — if I write it out I would get it wrong, certainly. We found that μ^2 was a negative infinite for values a little above the frequency of the highest critical period, or any other critical period. What does $\mu^2 = -\infty$ mean? It corresponds to a total reflection. Put "μ^2 is negative" into your analytical formula and you find the case in which vibrations cannot be propagated. We want a mechanical illustration of that. Do it by taking two heavy stretched cords connected by slight elastic bands — or rather take one stretched cord to show transverse vibrations, connected by very fine elastic bands with fixed points, and you will find that you cannot get a wave to go along it at all above a certain frequency, just as we cannot get a wave to go along this wave machine above a certain frequency but for a different reason, and in a different way. Put just work that out — it will take about three-quarters of an hour to do it nicely — and think of the interpretation of μ^2 negative: it will correspond to the case in which waves cannot get into the medium at all, and we have total reflection. We find an imaginary symbol introduced in the kind of solution we are familiar with. The corresponding kind of real symbols would express the thing. The use of the imaginary symbol for explaining the ordinary total internal reflection is perfectly straight-forward. It used to be made a very difficult thing; but now everybody knows what mathematicians were puzzled over 40 or 50 years ago, and that is the interpretation of a true dynamical formula whenever an imaginary symbol comes into it. You know that perfectly well. Green took that up and made it clear. Green was the first, I think, to give the total internal reflection of glass and so on. Precisely the same kind of analysis that gives you total internal reflection at very oblique incidences gives you total reflection even at direct incidences for certain frequencies a little above any of the critical periods. That agrees, I believe with observations. That ought to be the case with metals, although there are observations that go against the totality of the reflection; but if you look at appearances, it seems as if there ought to be total reflection. Silver is a shining instance; silver is total reflection all over. The molecular explanation of that property of silver would be simply that the highest mode, the shrillest mode of vibration of the molecules with which silver loads the luminiferous ether is graver than the mode of the gravest light or radiant heat that we have ever had reflected from silver. That is all. Is it improbable that the shortest period of the molecule in silver may not be greater than the twenty million-millionths of a second - is it not very probable that the quickest mode of vibration of the molecule in such a heavy body, a body of such high specific gravity as silver, may be at least 20 times as long as in the molecule of sodium. That is all that is assumed; surely that is probably enough.

But now, what if you get a little light through— take a piece of silver whose thickness is less than the wave length and some light will get through. I have not worked this theory out, but I hope to do so in going home so that you may have it in the report.*

* See Appendix.

We shall find no doubt that the light will get through that faster than in the luminiferous ether. Take gold leaf, say, of the thickness of half a wave length, or a quarter of a wave length — I have a specimen of such a leaf here, given to me by Prof. Trowbridge, I have an interest for some of you to see this specimen. Quincke has experimented upon very thin pieces of metal and has found that light passes through them with an acceleration. These are rather interesting experiments with gold of thickness engraved upon them of about the tenth or twentieth of a wave length. I am sorry we have not time to study them. I would have liked to have brought them before you all.

Suppose we have not μ^2 negative with total reflection, but μ^2 less than unity. First we have $\mu^2 = 0$, and then going on up to unity. In the positions for vibrations corresponding to μ^2 between 0 and 1 which is for periods a little shorter than a critical period, we should have acceleration in the substance, a velocity of propagation greater than in the luminiferous ether. If we had an hour to more carefully study the quantities concerned in the absorption of light by, for instance, sodium vapor, we should arrive at some very curious and interesting conclusions and thoughts. I am afraid we must leave it but think of a sodium flame in a hollow space in the interior of a glass globe, provided properly with air, with sodium vapor filling the globe so as to literally extinguish the flame in all directions. All the light that comes from that flame is absorbed into the sodium vapor. Think of the energy thus laid up, and you will get some very instructive lessons.

I will leave this sheet of to-day in your hands; it speaks for itself. You all understand that there must be continually work done in sending a wave in one direction. Take any portion of the wave front, and work must be done by the medium on one side of the wave front upon the medium on the other side to the extent exactly equal to the energy transmitted into the space beyond. If, then, the thing that is transmitted into the space is a succession of waves beginning abruptly and then perfectly regular and continuous — a succession of waves representing an arbitrary function, if you like — then the work done by the plane of the wave front per period must be equal to the sum of the kinetic and potential energies of the medium per wave length. That is the case in our ordinary formulas when we have $v^2 = \frac{k}{\rho}$, as we verified the other day. Now I call attention to this that when the medium is loaded with molecules the work done by the wave front exceeds the work done in the ether itself by the amount written down in this last formula. That is the amount of work then that goes to give energy to the attached molecules. It depends upon the spring arrangements, periods and so on, whether the energy taken by the molecules is not much greater than, or somewhat less then, or enormously greater than the energy in the elastic medium itself. When you come to the question of absorption bands, etc., the molecules will take thousands or perhaps millions of times as much energy as the energy of elastic action and motion in the ether itself. It is to prepare the way for this sort of thought that this paper is put in your hands. I think it sufficiently prepares the way. Suppose, for example, the energy of the molecules is two or three times the energy of the medium. Then it is perfectly clear that a succession of waves would go on advancing into the medium uniformly. The motion must be got up gradually. The result will be that if you commence a source of light and continue it quite constant for a length of time, there will be a gradual change in the first thousand, or the first hundred thousand, or the first million waves, but after a certain time it will be simply periodic. That will be the difference of circumstances from the circumstances we have to consider in the plane theory without attacked molecules, or with a homogeneous medium in which the work done by the wave front per period is equal to the energy per wave length and in which we advance without change of form of a single wave or group of waves. I am afraid the thing is very imperfect, but it is a most practical and important subject that we have to think of, such as it is.

I want you to look at this drawing of Langley's. This is a thing that is most important. Here are relations of wave lengths and refrangibilities, but this is the thing I want you to see. We are all familiar with that drawing. There is the thing we know so well that Herschel worked out showing where we have the maximum heat in the solar spectrum. Here again is the energy of a Leslie cube — a cube of hot water. There is the maximum, way down in 37 of the scale. It is most important to see the wave length corresponding to the maximum energy in the spectrum of a Leslie cube and to compare it with that of the Solar spectrum.

I am exceedingly sorry that our 21 coefficients are to be scattered, but though scattered far and wide, I hope we will still be coefficients working together for the great cause we are all so much interested in. I would be most happy to look forward to another conference and the one damper to that happiness is that this is now to end and we shall be compelled to look forward for a time. I hope only for a time and that we shall all meet again in some such way. I would say to those whose homes are on this side of the Atlantic, come on the other side and I will welcome you heartily and we may have more conferences. Whether we have such a conference on this side or on the other side of the Atlantic again, it will be a thing to look forward to, as this is looked back upon, as one of the most precious incidents I can possibly have. I suppose we must say farewell.

[Sir Wm. Thomson's allusion to the 21 coefficients will be explained by the following humorous poem read at a dinner party of the previous day, which was given to Sir William Thomson and the physicists in attendence upon his lectures, by President Gilman of the John's Hopkins University. The author is Prof. G. Forbes, of London, England.

The Lament of the 21 Coefficients in Parting from each other and from their esteemed Molecule.

An aeolotropic molecule was looking at the view
Surrounded by his coefficients, twenty-one or two,
And wondering whether he could make a sky of azure blue
With platitatic a b c and thlipsinomic 2.

They looked like sand upon the shore with waves upon the sea
But the waves were all too wilfull and determined to be free,
And in spite of n's rigidity they never could agree
In becoming quite subservient to the thlipsinomic P

Then web-like coefficients and a loaded molecule
With a noble wiggler at their head worked hard as
Naughton's mills,
But the waves all laughed and said a wiggler thinking
he could rule
A wave was nothing better than a sidelong normal fool

So the coefficients sighed and gave a last tangential skew
And a shook hands with &c and S and P and U,
And with a tear they parted, but they said they would be
true
To their much beloved wiggler and to thlipsinomic Q.

Signed (g. f.) a cross-coefficient
now annulled

President Gilman passed favorable verdict upon the versification, Sir Wm. Thomson said the mathematics seemed all right, and the coefficents unanimously concurred in the sentiments expressed. I therefore consider its insertion justifiable even in a more solemn and heavy scientific work than this purports to be.
H.]

— APPENDIX. —

Improved Gyrostatic Molecule.*

The efficiency of the gyrostatic molecule described in my lecture of 16th Oct., is obviously in simple proportion to the amount of moment of inertia per unit volume of the medium: this is clear on the supposition that the axes of all the molecules are parallel, and their rotations in the same direction. When the axes are turned in all directions, the sum of components of moments of momentum round three axes at right angles to one another, may be first taken, and their resultant in the usual manner of dealing with problems of moment of momentum. It is to the amount of this resultant moment of momentum per unit volume, that the required efficiency is proportional, whatever be the distribution of axes of the molecules through the medium. Farther it is easily proved that the rate of rotation of the plane of a distortional wave advancing through the medium per unit of distance travelled is, for different directions of the wave-normal, proportional to the cosine of its inclination to the direction of the resultant axis, determined in the manner just described. With these understandings, it will be convenient for the sake of simplicity, to deal particularly with the extreme case of the axes of all the molecules parallel, and their rotations in the same direction: also to suppose them all equal and similar. Let __a__ be the distance between the pivotted ends of the flywheels of each molecule (or the diameter of the spherical sheath, in=

*[Preliminary regarding molecule of Oct. 16, added Nov. 1st, 1884.]

agined in the little diagram of Oct. 16). Let $\underline{k}$ be the radius of gyration of the flywheel, and let $\underline{m}$ be the sum of the masses of all the flywheels, distributed through a volume of $8\pi^3$ of the ether. To admit of definite calculation, we must (as before in respect to our compound spring molecules) suppose the sum of the volumes of the spaces occupied by the sheaths of the molecules, to be infinitely small in comparison with the volume of the space filled with the homogeneous ether around them. It is easy to prove that the equations for wave motion, with wave front perpendicular to the axes of the rotations, are

$$\left.\begin{aligned} \frac{\rho}{4\pi^2}\,\frac{d^2\eta}{dt^2} &= \frac{\ell}{4\pi^2}\cdot\frac{d^2\eta}{dx^2} + \frac{mk^2\gamma}{8\pi^3}\cdot\frac{d^3\zeta}{dt\,dx^2} \\ \frac{\rho}{4\pi^2}\,\frac{d^2\zeta}{dt^2} &= \frac{\ell}{4\pi^2}\,\frac{d^2\zeta}{dx^2} - \frac{mk^2\gamma}{8\pi^3}\cdot\frac{d^3\eta}{dt\,dx^2} \end{aligned}\right\} (1);$$

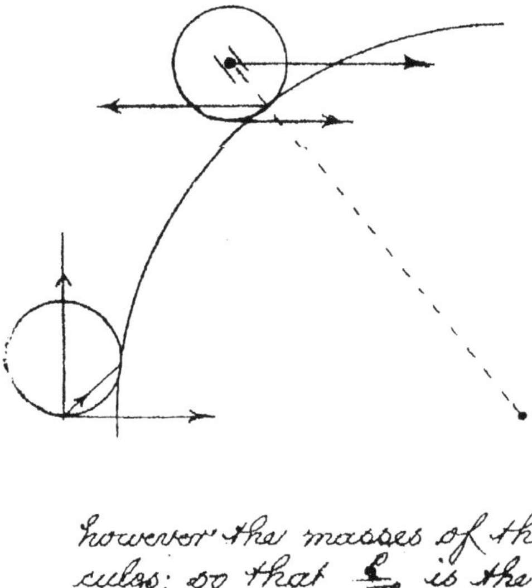

where $\underline{x}$ denotes distances from a fixed plane parallel to the wave front, and η, ζ, the components of displacement parallel to two fixed lines at right angles to one another in that plane. As previously, $\frac{\ell}{4\pi^2}$ denotes the rigidity of the ether, and $\frac{\rho}{4\pi^2}$ its density; including now however the masses of the sheaths and gyrostatic molecules; so that $\frac{\rho}{4\pi^2}$ is the average density of the whole material medium and imbedded molecules.

The most convenient way of dealing with these equations, is to apply them at once to investigate circularly polarized light. For this purpose let

$$\left.\begin{aligned} \eta &= \sin\,2\pi\left(\frac{x}{\lambda} - \frac{t}{T}\right) \\ \zeta &= -\cos\,2\pi\left(\frac{x}{\lambda} - \frac{t}{T}\right) \end{aligned}\right\} \qquad (2)$$

With this assumption, either of equations (1), gives

$$\frac{\rho}{T^2} = \frac{\ell}{\lambda^2} + \frac{mk^2\gamma}{T\lambda^2} \quad \ldots\ldots \quad (3);$$

hence
$$\frac{\lambda^2}{T^2} = \frac{\ell}{\rho} + \frac{mk^2\gamma}{\rho T} = \left(1 + \frac{mk^2\gamma}{\ell T}\right)\frac{\ell}{\rho} \quad \ldots \quad (4),$$

and therefore very approximately

$$\frac{\lambda}{T} = \left(1 + \frac{1}{2}\frac{mk^2\gamma}{\ell T}\right)\sqrt{\frac{\ell}{\rho}} \quad \ldots\ldots \quad (5).$$

Similarily of λ' denote the wave length for circularily polarized light, with orbital motions in the opposite direction to that expressed by equations (2), we find

$$\frac{\lambda'}{T} = \left(1 - \frac{1}{2}\frac{mk^2\gamma}{\ell T}\right)\sqrt{\frac{\ell}{\rho}} \quad \ldots\ldots \quad (6);$$

and instead of (2), we may take for this case

$$\left.\begin{aligned} \xi' &= \sin 2\pi\left(\frac{x}{\lambda'} - \frac{t}{T}\right) \\ \eta' &= \cos 2\pi\left(\frac{x}{\lambda'} - \frac{t}{T}\right) \end{aligned}\right\} \quad \ldots\ldots \quad (7).$$

The resultant (ξ'', η'') of the motions (2) and (7) superimposed is expressed by

$$\left.\begin{aligned} \xi'' &= \xi + \xi' = \sin 2\pi\left(\frac{x}{\lambda} - \frac{t}{T}\right) + \sin 2\pi\left(\frac{x}{\lambda'} - \frac{t}{T}\right) \\ \eta'' &= \eta + \eta' = \cos 2\pi\left(\frac{x}{\lambda} - \frac{t}{T}\right) - \cos 2\pi\left(\frac{x}{\lambda'} - \frac{t}{T}\right) \end{aligned}\right\} \ldots \quad (8).$$

Put now

$$\frac{1}{\lambda} = \frac{1}{\ell} - \frac{1}{a}, \quad \frac{1}{\lambda'} = \frac{1}{\ell} + \frac{1}{a} \quad \ldots\ldots \quad (9),$$

we find from (8)

$$\left.\begin{aligned} \xi'' &= 2\cos\frac{2\pi x}{a}\,\sin 2\pi\left(\frac{x}{\ell} - \frac{t}{T}\right) \\ \eta'' &= 2\sin\frac{2\pi}{a}\,\sin 2\pi\left(\frac{x}{\ell} - \frac{t}{T}\right) \end{aligned}\right\} \quad \ldots\ldots \quad (10).$$

These express the motion in a wave of transverse rectilinear vibrations of which the velocity of propagation is $\frac{\ell}{T} = \sqrt{\frac{e}{\rho}}$ (call this v) . . . . (11), and in which the direction of the vibrations is constant in every part of the medium, but turns round the direction of propagation, at the rate of one round per distance equal to a, of which the value, found from (9), (5) and (6), is

$$\left.\begin{aligned} a &= \left\{\frac{1}{2}\left(\frac{1}{\lambda} - \frac{1}{\lambda'}\right)\right\}^{-1} = \frac{\lambda\lambda'}{\frac{1}{2}(\lambda - \lambda')} \\ &= \frac{e\sqrt{\frac{\ell}{\rho}}}{mk^2\gamma}\,T^2 = \frac{\ell}{mk^2\gamma}\,vT^2 \end{aligned}\right\} \quad \ldots\ldots \quad (12)$$

Thus we see that the efficiency in rotative effect on the plane of polarization is equal to $m k^2 \gamma / c$. Suppose now the molecules to be made smaller and smaller.

(Continued on Page 320)

[In the lectures, $\frac{c}{4\pi}$ denotes the rigidity of the ether, I having mistaken the c in my notes for an l. I took it to be the same in the manuscript of the above and in handing it to the copyist instructed him to make it more plainly an l. In reading the proof, however, it seems to me that formula (9) introduces a new letter. The three letters l, e, c are very confusing in manuscript. Witness the following tracings from (1) and (9): $\frac{c}{4\pi^2}$, $\frac{1}{c}$.

The traced diagram two pages back was found upon the back of the manuscript page opposite without reference marks.

I have received the following correction from Sir Wm Thomson: "Please omit 'I do not find it quite &c.' and 'For instance Lord Rayleigh &c' (at top of p. 16) I think I found that I had misunderstood or misremembered one sentence of Lord Rayleigh's, and that what he said on this particular point was quite unobjectionable."

I infer (from a marginal note by W.T.) the following from Lord Rayleigh on Double Refraction is the sentence referred to:

"Fresnel and Green were inconsistent. The latter has given two rigorous theories of double refraction which differ from one another in important points, but agree in this, that neither of them can be reconciled with his explanation of reflection; for both assume that the forces which resist displacement within a crystal vary according to the direction of displacement. Precisely the same remark applies to investigations of Cauchy." H.T.

To test the molecular hypothesis for the reflection of light at the surfaces of metals, and the transmission of light through thin metal foils.

I. Metallic Reflection

(1) Vibrations perpendicular to the plane of incidence

(2) Vibrations in the plane of incidence

Notation and explanations as in the leaf of notes for lecture of Oct. 16th.

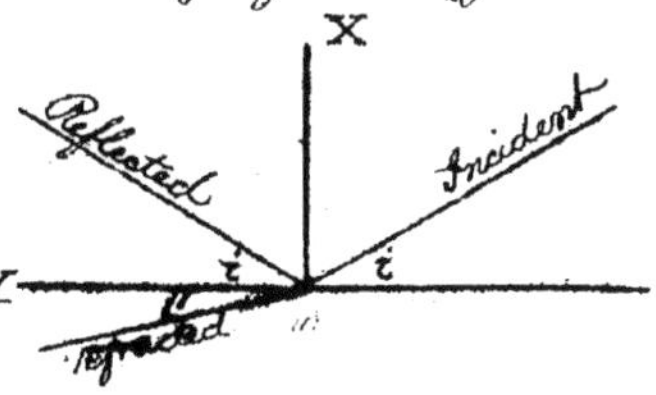

Adopting now the suppositions of incompressibility we have $\mathfrak{b} = b$; and the equations become

$$\left.\begin{cases} \psi = A\varepsilon^{\iota(ax+by+\omega t)} + A_{,}\varepsilon^{\iota(-ax+by+\omega t)} \\ \phi = B\varepsilon^{-bx+\iota(by+\omega t)} \end{cases}\right\} x \text{ positive}$$

for upper medium; and for lower medium

$$\psi = \varepsilon^{\iota(a'x+by+\omega t)}; \quad \phi = B'\varepsilon^{bx+\iota(by+\omega t)} \quad (x \text{ negative})$$

Then, $\xi \left(= \dfrac{d\phi}{dx} + \dfrac{d\psi}{dy}\right)$ equated for upper and lower mediums gives $-Bb + \iota b(A + A_{,}) = B'b + \iota b \ldots$

Then $\eta \left(= \dfrac{d\phi}{dx} + \dfrac{d\psi}{dy}\right)$ equated for upper and lower mediums " $\iota Bb - \iota a(A - A_{,}) = \iota B'b - \iota a' \ldots$

These yield $A + A_{,} = 1 - \iota(B + B')$; $A - A_{,} = \dfrac{a'}{a} + \dfrac{b}{a}(B - B') \quad \ldots (1)$

and so, conveniently, the problem is reduced to the determination of the interfacial wave (B, B').

The other two equations are found as follows.

$P\left(= p^{*} + 2n\dfrac{d\xi}{dx}\right)$ equated for upper and lower mediums, gives

$$n\{(a^2 - b^2)B + 2ab(A - A_{,})\} = n'\{(a'^2 - b^2)B' + 2a'b\} \quad \ldots (P)$$

$U\left[= n\left(\dfrac{d\xi}{dy} + \dfrac{d\eta}{dx}\right)\right]$ equated for upper and lower mediums, gives

$$n\{-2\iota b^2 B + (a^2 - b^2)(A + A_{,})\} = n'\{2\iota b^2 B' + a'^2 - b^2\} \quad \ldots (U)$$

Eliminating $A - A_{,}$ and $A + A_{,}$ from these by (1) we find

$$\left.\begin{aligned} &n(a^2+b^2)B - \{n'(a'^2+b^2) - 2(n'-n)b^2\}B' = 2(n'-n)a'b \\ \text{and } &n(a^2+b^2)B + \{n(a^2+b^2) + 2(n'-n)b^2\}B' = \iota[n'a'^2 - na^2 - (n'-n)b^2] \end{aligned}\right\} \ldots (2)$$

These two equations determine B and B', and the results in (1) give A and $A_{,}$, so completing the solution of the

---

* On the leaf of notes of Oct 16th, we saw that $p = -\rho\omega^2\phi = -n(a^2+b^2)\phi$ in upper medium and the same with accents for the lower medium.

problem. It is interesting and important, not only for the wave theory of light, but for the dynamics of elastic solids, to work out explicitly and to thoroughly interpret this solution without any restriction as to the rigidities (n, n') or the densities (ρ, ρ'). Meantime for reasons already considered we shall suppose $n = n'$, by which at this stage a great simplification is produced, reducing (2) to

$$\left.\begin{array}{l}(a^2+b^2)B-(a'^2+b^2)B'=0 \quad \cdots (3)\\ (a^2+b^2)(B+B')=\iota(a'^2-a^2) \quad \cdots (4)\end{array}\right\}(\text{case } n=n').$$

Now we have $a^2+b^2=\frac{4\pi^2}{\lambda^2}$, $a'^2+b^2=\frac{4\pi^2}{\lambda'^2}$ $\cdots\cdots$ (5)

if λ, λ' denote the wave length in the upper and lower mediums. Hence (3) gives

$$\frac{B}{\lambda^2}=\frac{B'}{\lambda'^2}=\frac{B+B'}{\lambda^2+\lambda'^2}=\frac{B-B'}{\lambda^2-\lambda'^2} \quad \cdots\cdots (6):$$

and (4) gives

$$B+B'=\iota\frac{\lambda^2-\lambda'^2}{\lambda'^2} \quad \cdots\cdots (7).$$

From this and (6) we find

$$B-B'=\iota\frac{(\lambda^2-\lambda'^2)^2}{\lambda'^2(\lambda^2+\lambda'^2)} \quad \cdots\cdots (8);$$

and this again with (6) gives

$$B=\iota\frac{\lambda^2(\lambda^2-\lambda'^2)}{\lambda'^2(\lambda^2+\lambda'^2)};\ B'=\iota\frac{\lambda^2-\lambda'^2}{\lambda^2+\lambda'^2} \quad \cdots\cdots (9).$$

Denote now by μ the index of refraction from the upper to the lower medium. We have

$$\frac{\lambda}{\lambda'}=\mu \quad \cdots\cdots (10)$$

Hence (7), (8) and (9) become

$$B+B'=\iota(\mu^2-1),\ B-B'=\iota\frac{(\mu^2-1)^2}{\mu^2+1} \quad \cdots (11),$$

and

$$B=\iota\frac{\mu^2(\mu^2-1)}{\mu^2+1},\ B'=\iota\frac{\mu^2-1}{\mu^2+1} \quad \cdots\cdots (12).$$

Remembering that

$$\left.\begin{array}{l}a=\frac{2\pi}{\lambda}\cos i,\ b=\frac{2\pi}{\lambda}\sin i;\ a'=\frac{2\pi}{\lambda'}\cos i',\ b=\frac{2\pi}{\lambda'}\sin i';\\ \frac{b}{a}=\tan i, \text{ and } \frac{a'}{a}=\mu\frac{\cos i'}{\cos i}=\frac{\tan i}{\tan i'}\end{array}\right\}(13),$$

and using (11) in (1) we find

$$A+A_1=\mu^2,\ A-A_1=\frac{\tan i}{\tan i'}+\iota\tan i\frac{(\mu^2-1)^2}{\mu^2+1} \quad \cdots (14).$$

Hence, finally, for our solution we have

$$\text{in upper medium}\left\{\begin{array}{l}\psi=\frac{1}{2}\left\{\mu^2+\frac{\tan i}{\tan i'}+\iota\tan i\frac{(\mu^2-1)^2}{\mu^2+1}\right\}\varepsilon^{\iota(ax+by+\omega t)}\\ \psi_1=\frac{1}{2}\left\{\mu^2-\frac{\tan i}{\tan i'}-\iota\tan i\frac{(\mu^2-1)^2}{\mu^2+1}\right\}\varepsilon^{\iota(-ax+by+\omega t)}\\ \varphi=\iota\frac{\mu^2(\mu^2-1)}{\mu^2+1}\varepsilon^{-bx+\iota(by+\omega t)}\end{array}\right\}\cdots (15)$$

$$\text{in lower medium}\left\{\begin{array}{l}\varphi = \iota\dfrac{\mu^2-1}{\mu^2+1}\,\varepsilon^{bx+\iota(by+\omega t)}\\ \psi = \varepsilon^{\iota(a'x+by+\omega t)}\\ \psi_1 = 0\end{array}\right\}\cdots\cdots(16).$$

To realize for the case of μ real, change ι into $-\iota$ and add the results to the preceding. Altering the notation correspondingly, to let ψ, φ, &c., denote real functions, we thus find:

$$\text{in upper medium}\left\{\begin{array}{l}\psi = \left(\mu^2+\dfrac{\tan i}{\tan i'}\right)\cos(ax+by+\omega t)-\dfrac{(\mu^2-1)^2}{\mu^2+1}\sin(ax+by+\omega t)\\ \psi = \left(\mu^2-\dfrac{\tan i}{\tan i'}\right)\cos(-ax+by+\omega t)+\dfrac{(\mu^2-1)^2}{\mu^2+1}\sin(-ax+by+\omega t)\\ \varphi = -\dfrac{\mu^2(\mu^2-1)}{\mu^2+1}\,\varepsilon^{-bx}\sin(by+\omega t)\end{array}\right\}(17),$$

$$\text{and in lower medium}\left\{\begin{array}{l}\varphi = \dfrac{\mu^2-1}{\mu^2+1}\,\varepsilon^{bx}\sin(by+\omega t)\\ \psi = 2\cos(a'x+by+\omega t);\ \psi_1 = 0\end{array}\right\}\cdots\cdots(18).$$

Let now ω be the resultant displacement of any part of either medium at a distance from the interface large in comparison with the wave length. The interfacial wave, φ, contributes nothing sensible towards this resultant and we have, as is easily seen (from (··), (··), above

$$\omega = \eta\sec i = -\frac{d\psi}{dx}\sec i \cdots\cdots (19).$$

Before using this, reduce ψ and ψ_1 for the upper medium to the normal simple-harmonic form $R\cos(q+e)$ and $R_1\cos(q_1-e)$, by the notation

$$\left.\begin{array}{l}\tan e = \dfrac{(\mu^2-1)^2}{\mu^2+1}\Big/\left(\mu^2+\dfrac{\tan i}{\tan i'}\right);\ R = \left(\mu^2+\dfrac{\tan i}{\tan i'}\right)\sec e\\ \tan e_1 = \dfrac{(\mu^2-1)^2}{\mu^2+1}\Big/\left(\mu^2-\dfrac{\tan i}{\tan i'}\right);\ R_1 = \left(\mu^2-\dfrac{\tan i}{\tan i'}\right)\sec e_1\end{array}\right\}\cdots\cdots(20).$$

Then by (19) and (13), we find

$$\left.\begin{array}{ll}\text{in upper medium}\left\{\begin{array}{l}\omega = \dfrac{2\pi}{\lambda}R\sin(ax+by+\omega t+e)\cdots\text{incident wave}\\ \omega_1 = \dfrac{2\pi}{\lambda}R_1\sin(-ax+by+\omega t-e_1)\cdots\text{reflected wave}\end{array}\right\} & \\ \text{and in lower medium}\ \omega = 2\dfrac{2\pi}{\lambda}\sin(a'x+by+\omega t)\cdots\text{refracted wave} & \end{array}\right\}\cdots(21)$$

which agrees with Green's original solution. The formula which I quoted (Oct 17) from Lord Rayleigh comes immediately from it; as also his formula for retardation of phase which I had not time to quote. But our present affair is the case of $-\mu^2$ a real positive numeric, for which we must now realize the symbolic formulas (15), (16).

Because $\sin i'$ is now imaginary we may conveniently replace in (15) $\tan i / \tan i'$, by a'/a its value according to (13); and for a' as follows;

$$\left.\begin{array}{l} a'^2 + b^2 = -V^2(a^2+b^2), \text{ whence } \iota a' = h \\ \text{where } V^2 = -\mu^2;\ h = \{(V^2+1)b^2 + V^2a^2\}^{\frac{1}{2}} = \frac{2\pi}{\lambda}(V^2+\sin^2 i)^{\frac{1}{2}} = a(V^2\sec^2 i + \tan^2 i)^{\frac{1}{2}} \end{array}\right\} \quad (22)$$

Thus (15) and (16) become

$$\text{in the upper medium} \left\{\begin{array}{l} \psi = \frac{1}{2}\left\{-V^2 - \iota\frac{h}{a} - \iota\tan i\frac{(V^2+1)^2}{V^2-1}\right\}\varepsilon^{\iota(ax+by+\omega t)} \\ \psi_1 = \frac{1}{2}\left\{-V^2 + \iota\frac{h}{a} + \iota\tan i\frac{(V^2+1)^2}{V^2-1}\right\}\varepsilon^{\iota(-ax+by+\omega t)} \\ \varphi = -\iota\frac{V^2(V^2+1)}{V^2-1}\varepsilon^{-bx+\iota(y+\omega t)} \qquad [(by+\omega t)\iota] \end{array}\right\} \cdots (23);$$

$$\text{and in the lower medium} \left\{\begin{array}{l} \varphi = \iota\frac{V^2+1}{V^2-1}\varepsilon^{bx+\iota(by+\omega t)} \\ \psi = \varepsilon^{hx+\iota(by+\omega t)} \\ \psi_1 = 0 \end{array}\right\} \cdots\cdots (24)$$

Adding solutions for $\pm\iota$ and altering the notation of ψ and φ to let them denote real functions, we find

$$\text{in upper medium} \left\{\begin{array}{l} \psi = -V^2\cos(ax+by+\omega t) + \tan i\frac{(V^2+1)^2}{V^2-1}\sin(ax+by+\omega t) \\ \psi_1 = -V^2\cos(ax+by+\omega t) - \tan i\frac{(V^2+1)^2}{V^2-1}\sin(-ax+by+\omega t) \\ \varphi = \frac{V^2(V^2+1)}{V^2-1}\varepsilon^{-bx}\sin(y+\omega t) \qquad [(by+\omega t)\iota] \end{array}\right\} \cdots (25)$$

$$\text{and in lower medium} \left\{\begin{array}{l} \varphi = -\frac{V^2+1}{V^2-1}\varepsilon^{bx}\sin(y+\omega t) \\ \psi = 2\varepsilon^{hx}\cos(by+\omega t),\quad \psi_1 = 0 \end{array}\right\} \cdots\cdots\cdots (26).$$

To reduce to normal simple harmonic form, put

$$\left[\frac{h}{a} + \tan i\frac{(V^2+1)^2}{V^2-1}\right] V^2\tan f,\quad S = V^2\sec f \quad \cdots\cdots (27);$$

and we find

$$\text{in upper medium} \left\{\begin{array}{l} \psi = -S\cos(ax+by+\omega t+f) \\ \psi_1 = -S\cos(-ax+by+\omega t-f) \end{array}\right\} \cdots\cdots (28)$$

Hence, as above, we find for the resultant displacements of parts of the upper medium at distances from the interface great in comparison with the wave length,

$$\text{in the upper medium} \left\{\begin{array}{ll} \omega = -\frac{2\pi}{\lambda}S\sin(ax+by+\omega t+f) \cdots & \text{incident wave} \\ \omega = \frac{2\pi}{\lambda}S\sin(-ax+by+\omega t-f) \cdots & \text{reflected wave} \end{array}\right\} (29).$$

Having thus completed the work with the simplification of $n = n'$ which, following Green, we introduced into equations (3), and have kept in all up to equations (29), it is worth while now to take the general solution without this simplification which I have worked out, in the first place for the sake of endeavoring to judge whether or not there is advantage to be gained for the wave theory of light by supposing the effective rigidities different in different mediums, and in the second place because the general solution is in itself interesting in the theory of elastic solids. Going back to equations (2) put for brevity

$$\left.\begin{array}{l} a^2+b^2=1;\ \frac{n'}{n}=r;\ \text{and}\ \mu^2=\frac{n\rho'}{n'\rho};\ \text{which makes}\ \sqrt{(a'^2+b^2)}=\mu; \\ \text{and therefore}\ a'=\sqrt{(\mu^2-b^2)};\ \ a=\cos i;\ \ b=\sin i. \end{array}\right\} \quad (30)$$

Put also $2(r-1)b^2 = u$ . . . . . . (31)

we may now write equations (2) as follows:—

$$\left.\begin{array}{l} B-(r\mu^2-u)B' = \frac{a'}{b}u \\ \text{and}\ B+(1+u)B' = \iota(r\mu^2-1-u) \end{array}\right\} \ . \ . \ . \quad (32).$$

From these we find

$$\left.\begin{array}{l} (1+r\mu^2)B' = \iota(r\mu^2-u-1)-\frac{a'}{b}u \\ (1+r\mu^2)B = \iota(r\mu^2-u)(r\mu^2-u-1)+\frac{a'}{b}u(1+u) \end{array}\right\} \ . \ . \quad (33),$$

and using these (33) in (1) above we find

$$\left.\begin{array}{l} (1+r\mu^2)(A+A_1) = r\mu^2+(r\mu^2-u)^2-\iota\frac{a'}{b}u^2 \\ (1+r\mu^2)(A-A_1) = \frac{a'}{a}\left\{r\mu^2+(1+u)^2+\iota\frac{b}{a}(r\mu^2-u-1)^2\right\} \end{array}\right\} \ . \ . \ . \quad (34).$$

Abbreviate by putting $r\mu^2 = D$ . . . . (35);

by which D shall denote the ratio of density of the lower medium, to density of the upper medium. Thus finally

$$\left.\begin{array}{l} 2(1+D)A = D+(D-u)^2+\frac{a'}{a}\left\{D+(1+u)^2\right\}+\iota\frac{b}{a}\left\{(D-u-1)^2-\frac{aa'}{b^2}u^2\right\} \\ 2(1+D)A_1 = D+(D-u)^2-\frac{a'}{a}\left\{D+(1+u)^2\right\}-\iota\frac{b}{a}\left\{(D-u-1)^2+\frac{aa'}{b^2}u^2\right\} \end{array}\right\} \quad (36),$$

which is the final solution, in form convenient for being realized in either of the cases μ^2 real positive, or μ^2 real negative. The realized forms for the case of μ real (D real and positive) are obvious from the equations, and need not be written down here. In the the case of $r=1$, u vanishes and we fall back on equations (14) with their consequences (20), (21) alone as a particular case of these (36).

In the particular case of $r = \frac{1}{\mu^2}$, which makes the densities equal in the two mediums, we ought, as we shall see below, to find a find a result not differing greatly from Fresnel's "sine-formula," if MacCullagh's admirable but seductive explanation of Fresnel's tangent-formula, by vibrations perpendicular to the plane of the three rays, were correct. How wildly wide of agreement with Fresnel's sine-formula or with anything in nature respecting the reflection of light, is the supposition of equal densities and unequal rigidities in the two mediums, was discovered by Lorenz and Rayleigh partly from examination of the particular case of $\mu - 1$ infitely small and vibrations in the plane of the three rays, in the problem of reflection and refraction at a plane surface, and partly by Rayleigh's dynamics of the blue sky. Our general solution (36) agrees of course for the particular case referred to with the Lorenz's and Rayleigh's; and it serves to accentuate their important conclusion by showing equally wild results for all values of μ. I have worked it out for several angles of incidence, for the cases of $\mu = 1.225$, and $\mu = 1.5$ and have found curiously interesting results, which I need not give here as they have no importance for the wave theory of light unless as confirming what scarcely needed confirmation.

Our general solution (36) is also useful in dispelling the idea that, if Rood's experimental verification of Fresnel's formula $\left(\frac{\mu - 1}{\mu + 1}\right)^2$ for the intensity of light reflected nearly at right angles from transparent bodies, did not bar the way, we might, by giving r some value differing largely from either 1 or $\frac{1}{\mu^2}$, get something desirable whether for light polarized in the planes of the three rays or perpendicularly to it, out of the case of vibrations *in* the plane of the three rays. We see in fact $r=1$, or $r=1$, is the only supposition; that gives any approach to agreement to anything in nature respecting the reflection or refraction of light in transparent mediums. But, alas, we see also that the approach which the supposition $r=1$ (Green's theory) gives to explanation of the known phenomena of polarization is sadly distant, and that no either small or large change from the exact value 1, for r, can better it.

For our immediate purpose, of trying to see something of dynamical explanation for metallic reflection, let us realize (36) for μ^2 real and negative. As in (22). (29) above, with our present abbreviations $a^2+b^2=1$, we now have

$$\left.\begin{array}{l} \mu^2=-\nu^2;\ h=(\nu^2+b^2)^{\frac{1}{2}};\ a'=-2h;\ u=2(r-1)b^2;\ r=\frac{n'}{n}; \\ a=\cos i;\ b=\sin i:\ \text{put also } -D=\chi \end{array}\right\}\ldots(37)$$

so that, as the effective density of the lower medium is now negative, χ is the corresponding positive ratio to the density in the upper medium.

Thus equations (36) and (33) become

$$\left.\begin{array}{l} 2(1-\chi)A=-\chi+(\chi+u)^2-\frac{h}{b}u^2+\iota\left\{\frac{h}{a}[\chi-(1+u)^2]+\frac{b}{a}(\chi+u+1)^2\right\} \\ 2(1-\chi)A_1=-\chi+(\chi+u)^2-\frac{h}{b}u^2-\iota\left\{\frac{h}{a}[\chi-(1+u)^2]+\frac{b}{a}(\chi+u+1)^2\right\} \\ (1-\chi)B=\iota\left\{(\chi+u)(\chi+u+1)-\frac{h}{b}u(1+u)\right\} \\ (1-\chi)B'=-\iota\left(\chi+u+1-\frac{h}{b}u\right) \end{array}\right\}\ldots(38)$$

To realize as usual, put

$$\left.\begin{array}{l} \frac{h}{a}[\chi-(1+u)^2]+\frac{b}{a}(\chi+u+1)^2=H \\ -\chi+(\chi+u)^2-\frac{h}{b}u^2=K \end{array}\right\}\ldots\ldots(39);$$

and, (modifying the ψ, φ notation suitably for realization,) we have

$$\left.\begin{array}{ll} \text{upper medium} & \left\{\begin{array}{l} 2(1-\chi)\psi=H\sin(ax+by+\omega t)+K\cos(ax+by+\omega t) \\ 2(1-\chi)\psi_1=H\sin(-ax+by+\omega t)-K\cos(-ax+by+\omega t) \\ (1-\chi)\varphi=[(\chi+u)(\chi+u+1)-\frac{h}{b}u(1+u)]\varepsilon^{-bx}\cos(by+\omega t) \end{array}\right. \\ \text{lower medium} & \left\{\begin{array}{l} (1-\chi)\varphi=-(\chi+u+1-\frac{h}{b}u)\varepsilon^{bx}\cos(by+\omega t) \\ \psi=\varepsilon^{hx}\sin(by+\omega t) \end{array}\right. \end{array}\right\}\ldots(40)$$

Put now

$$\left.\begin{array}{l} K = R\sin f,\ H = R\cos f;\ \text{whence } \tan f = \frac{K}{H},\ R = \sqrt{(H^2+K^2)} \\ \text{which gives } 2(1-\chi)\psi = R\sin(ax+by+\omega t+f) \\ \qquad\qquad 2(1-\chi)\psi_1 = R\sin(-ax+by+\omega t-f) \end{array}\right\} \cdots (41);$$

and, by the fundamental equations, preceding (1) above, we have for the components of displacement:—

$$\left.\begin{array}{ll} \text{Incident wave} & \begin{cases} 2(1-\chi)\eta = -a\,R\cos(ax+by+\omega t+f); \\ \xi = -\frac{b}{a}\,\eta; \end{cases} \\ \text{Reflected wave} & \begin{cases} 2(1-\chi)\eta = a\,R\cos(-ax+by+\omega t-f); \\ \xi = \frac{b}{a}\,\eta; \end{cases} \\ \left.\begin{array}{l}\text{Interfacial wave in either medium} \\ \text{with proper values of } \varphi \text{ from (40)}\end{array}\right\} & \xi = \frac{d\varphi}{dx},\ \eta = \frac{d\varphi}{dy}; \\ \left.\begin{array}{l}\text{Motion of lower medium} \\ \text{in lieu of refracted wave}\end{array}\right\} & \begin{array}{l} \eta = -h\varepsilon^{hx}\sin(by+\omega t); \\ \xi = b\varepsilon^{hx}\cos(by+\omega t). \end{array} \end{array}\right\} \cdots (42)$$

Before interpreting this result let us find the corresponding result [(56) below] for the much easier case, of vibrations perpendicular to the plane of polarization. The differential equations for the upper and lower mediums respectively are,

$$\left.\begin{array}{l} \text{for upper medium } (x \text{ positive})\ \rho\frac{d^2\zeta}{dx^2} = n\left(\frac{d^2\zeta}{dx^2} + \frac{d^2\zeta}{dy^2}\right) \\ \text{and for lower medium } (x \text{ negative})\ \rho'\frac{d^2\zeta}{dx^2} = n'\left(\frac{d^2\zeta}{dx^2} + \frac{d^2\zeta}{dy^2}\right) \end{array}\right\} \cdots (43)$$

The stresses for this case of motion clearly involve solely tangential force in the plane of the wave front, and perpendicular to (Y X) (the plane of the diagram).

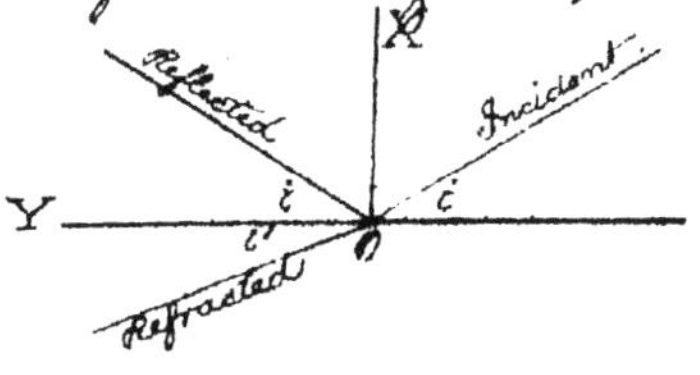

The component of this stress in any plane parallel to the interface between the two mediums, being the T of our general notation, is as follows for the two mediums

$$\left.\begin{array}{ll} (\text{upper}) & T = n\frac{d\zeta}{dx} \\ (\text{lower}) & T = n'\frac{d\zeta}{dx} \end{array}\right\} \quad . \quad . \quad . \quad . \quad . \quad (44).$$

Our solution in the case of vibration in the plane

of the three rays, might have been worked out from the beginning for a wave represented by any arbitrary periodic function, but it was more convenient for the ordinary analytic method of imaginaries which we used to work it out in the first place for exponentials and simple harmonic functions. But the fact that the resulting laws of refraction and reflection do not involve the wave length, suffice to prove them true for waves or pulses represented by arbitrary periodic or non-periodic functions. In the present case there is no advantage in point of simplicity or convenience in expressing our work in terms of exponentials or simple harmonic formulas. Let us follow Green therefore in taking the arbitrary solution as follows:

$$\left.\begin{array}{ll} \text{upper medium} & \xi = Af(ax+by+\omega t) + Bf(-ax+by+\omega t) \\ & \qquad \text{incident wave} \qquad \text{reflected wave} \\ \text{lower medium} & \xi = f(a'x+by+\omega t) \\ & \qquad \text{refracted wave} \end{array}\right\} (45);$$

which satisfies equation (43) provided

$$\left.\begin{array}{l} \rho\omega^2 = n(a^2+b^2) \\ \rho'\omega^2 = n'(a'^2+b^2) \end{array}\right\} (46).$$

The conditions to be satisfied at the interface, being equality of ξ on the two sides of it, and equality of T on the two sides of it, are expressed by the following equations;

$$\left.\begin{array}{l} A+B=1 \\ na(A-B)=n'a' \end{array}\right\} (47);$$

by which we find

$$\left.\begin{array}{l} A=\frac{1}{2}\left(\frac{n'a'}{na}+1\right) \\ B=-\frac{1}{2}\left(\frac{n'a'}{na}-1\right) \end{array}\right\} (48).$$

Still denoting as before by i and i', the angles of refraction and incidence, and putting now

$$\left.\begin{array}{l} \sqrt{a^2+b^2} = c \qquad \sqrt{a'^2+b^2} = c' \\ \text{we have } c' = c\sqrt{\frac{n\rho'}{n'\rho}} = c\mu; \ \sin i' = \sin i/\mu \\ b = c \sin i = c' \sin i'; \\ a = c \cos i; \ a' = c' \cos i'; \ \frac{a'}{a} = \frac{\tan i}{\tan i'}, = c(\mu^2 - \sin^2 i)^{\frac{1}{2}} \end{array}\right\} \quad (49)$$

Using this in equation (48), we find

$$\frac{B}{A} = -\frac{n' \tan i - n \tan i'}{n' \tan i + n \tan i'} \quad . \quad . \quad . \quad . \quad (50),$$

whichch expresses the ratios of the amplitude of the refracted to the amplitude of the incident ray. For the particular case of $n = n'$, this gives

$$\text{Case I} \qquad \left.\begin{array}{l} \text{Case } n = n' \\ \frac{\rho'}{\rho} = \frac{\sin^2 i}{\sin^2 i'} = \mu^2 \end{array}\right\} \frac{B}{A} = -\frac{\sin(i-i')}{\sin(i+i')} \quad . \quad . \quad . \quad . \quad (51),$$

which is Fresnel's celebrated sine-formula. By squaring each member we have his expression for the ratio of the intensity of the reflected to the intensity of the incident light. The negative sign shews change of plane by half a period on the reflected ray relatively to the transmitted ray: of course there is no distinction in the circumstances between retardation and acceleration of half a period and therefore we cannot say which it is.

$$\text{Case II.} \quad \left.\begin{array}{l} \rho = \rho' \\ \frac{n}{n'} = \frac{\sin^2 i}{\sin^2 i'} = \mu^2 \end{array}\right\} \left.\begin{array}{l} \frac{B}{A} = \frac{\sin^2 i \tan i' - \sin^2 i' \tan i}{\sin^2 i \tan i' + \sin^2 i' \tan i} \\ \quad = \frac{\sin i \cos i - \sin i' \cos i'}{\sin i \cos i + \sin i' \cos i'} = \frac{\tan(i-i')}{\tan(i+i')} \end{array}\right\} (52)$$

The last member is Fresnel's celebrated tangent-formula, which he gives for vibrations <u>in</u> the plane of the three rays. The very curious result that this formula expresses rigorously the law of reflection for vibrations <u>perpendicular</u> to the plane of the three rays, in the case of equal densities and unequal rigidities in the two mediums, seems to have been first discovered by McCullagh. It is most tempting in respect to the explanation of polarization by reflection. It tempts us to suppose with McCullagh the line of vibration to be *on* the plane of polarization, because at angle of incidence = $\tan^{-1}\mu$, a wave of vibrations perpendicular to the plane of the three rays, gives rise to no reflected light, and is transmitted without loss of energy into the lower medium of our diagram. But if this were the case the law of reflection of a wave of vibrations in the plane of the three rays should agree with Fresnel's sine-formula, or at all events should not differ from it more than observation allows us to suppose that light polarized in the plane of the three rays, can in reality differ from that formula. But also, Lorenz and Rayleigh * have shown that instead of fulfilling Fresnel's sine-law, the reflected ray in a wave of vibrations *in* the plane of the three rays would vanish at angles of incidence equal respectively to one-quarter, and three quarters of a right angle, when the index of refraction from one medium to the other, differs little from unity ($\mu \doteq 1$). This they find by working out formulas equivalent to our equations (1) and (2) above for the case $\rho = \rho'$ and $\mu = \sqrt{\frac{n}{n'}}$. They, therefore, with a cogency of which the force is clearly irresistable, concluded that the difference of the velocity of light in different mediums cannot be due to the difference of effective rigidities with equal effective densities or with approximately equal effective densities of the vibrating substance in the two mediums, and that in polarized light the vibrations are perpendicular to the plane of polarization. Lorenz went farther and concluded not merely that the difference of velocity is not due to difference of effective rigidity but that it is wholly due to "difference of density "in all transparent uncrystalline substances."

---

*See the Hon. J. W. Strutt (now Lord Rayleigh) Phil. Mag. Aug. 1871.

Rayleigh accepting this conclusion, refused to limit it to uncrystalline substances: his words are, "Lorenz draws" "the conclusion that the elastic force of the ether is the" "same in all transparent uncrystalline substances as in" "vacuo and that the vibrations of light are performed" "normally to the plane of polarization. He might, I think" "have omitted the word uncrystalline."

I cannot myself quite admit this as a conclusion from the premises. I do not see that there is sufficient ground in any of the phenomena referred to by either Lorenz or Rayleigh, for inferring that the effective rigidity is exactly, or is even very approximately, equal in the two mediums. It might be for instance that the rigidity is greater in the denser medium, but not greater in the same proportion as the density. This would make the velocity of propagation less in the denser medium, and it would give another available constant besides the index of refraction (and another is imperatively needed) to account for the enormously great difference between the results of observation, and of Green's theory as expressed in equations (20) and (21) above, in respect to the law of reflection of light polarized perpendicularly to the plane of the three rays, that is to say light of which the vibrations are in the plane of the three rays. But any such difference of rigidity, to be sufficient to go any considerable way towards accounting for the prodigious discrepancy between observation and Green's theory, would cause the reflected light at approximately perpendicular incidence to be vastly greater than $\{\frac{\mu-1}{\mu+1}\}^2$ of the incident light, which Green's theory, on the supposition of equal rigidities, makes it. I know of no observations bearing upon this point except those of Prof. Rood of Columbia College, New York.* They also, make the

* "On the amount of Light transmitted by plates of polished Crown Glass at a perpendicular incidence" American Journal of Science and Arts, Vol. L, July 1870

agreement with the $\left\{\frac{\mu-1}{\mu+1}\right\}^2$ law exceedingly close for Crown Glass, and as Prof. Rood himself informed me, when I had the good fortune to see him in New York, immediately after the conclusion of my Lectures in Baltimore, that the unpublished observations on flint glass and quartz to which he referred in his paper confirm the same law for them also to a somewhat close degree of accuracy, notwithstanding the imperfection of adjustment to which he alludes, as the reason which caused him to withhold them from publication. It seems therefore that after all we must accept the conclusion of Lorenz and Rayleigh, that the rigidity of the luminiferous ether is equal, or is at all events very approximately equal in ordinary transparent solids. It remains however, for experimental examination to find whether or not the rigidity is also equal in transparent liquids and in extreme cases of transparent solids such as diamond ($\mu = 2.47$ to 2.75) and sulphuret of arsenic ($\mu = 2.454$), and I see no way of deciding the question except by photometric experiments such as Rood's.

On the meantime Jamin's beautiful discovery of what he calls positive and negative reflection * remains without dynamical explanation. It violates Cauchy's formulas, but they are empirical and not dynamical. They have great merit as empirical formulas, and no dynamical law being fulfilled by Cauchy's "theory" none is broken by the modifications which Jamin, and Quincke ** in pursuing similar investigations, have given to Cauchy's formula to cause them to agree with observation.

But metallic reflection is our present subject and

---

* Annales de Chemie et de Physique, Vol. 29, 1850, page 263.
** Annalen der Physik und Chemie Vol. 119, 1863, p. 368; Vol. 127, 1866, pp. 1 and 199; Vol. 128, 1866, pages 355 & 541; Vol. 129, 1866, pages 44 and 177.

therefore let us realize the solution (48) for the case of $\mu^2 = -V^2$, V real. Take now instead of (46) with its arbitrary function, f, the ordinary exponential imaginary formulas, thus:—

$$\left.\begin{array}{ll}\text{upper medium} & \xi = A\varepsilon^{\iota(ax+by+\omega t)} + B\varepsilon^{\iota(-ax+by+\omega t)} \\ \text{lower medium} & \xi = \varepsilon^{\iota(a'x+by+\omega t)}\end{array}\right\} \quad \ldots (53)$$

Looking to (49) we see that i' is now imaginary, but i remains real, and by this the expression for a' becomes

$$a' = -\iota c(V^2 + \sin^2 i)^{\frac{1}{2}} \quad \ldots \quad (54)$$

Eliminating a' by this and a by its expression in (49), (48) becomes

$$\left.\begin{array}{l} A = \frac{1}{2}\left\{1 - \iota r(V^2\sec^2 i + \tan^2 i)^{1/2}\right\} \\ B = \frac{1}{2}\left\{1 + \iota r(V^2\sec^2 i + \tan^2 i)^{1/2}\right\}\end{array}\right\}, \quad \ldots \quad (55)$$

where $r = \frac{n'}{n}$

To realize as usual put

$$\left.\begin{array}{l} \tan e = r(V^2\sec^2 i + \tan^2 i)^{\frac{1}{2}} \\ \text{and} \quad R = \frac{1}{2}\left\{1 + r^2(V^2\sec^2 i + \tan^2 i)\right\}; \\ \text{we find} \\ \text{incident ray } \xi = R\cos(ax + by + \omega t - e) \\ \text{reflected ray } \xi = R\cos(ax + by + \omega t + e) \\ \left.\begin{array}{l}\text{and (in lieu of refracted ray)} \\ \text{motion in lower medium}\end{array}\right\} \xi = \varepsilon^{xc(V^2+\sin^2 i)^{1/2}}\cos(by + \omega t)\end{array}\right\} (56)$$

When we return a little later to the molecular theory developed in the lectures, we shall see that for periods slightly less than any one of the critical periods ($\varkappa_1, \varkappa_2$ &c of our former notation) the value of μ^2 is negative, and that for a wide proportionate range, say from $T = \varkappa_1$ to $T = \varkappa_1/N$, where N denotes some large numeric, we may have μ^2 negative, diminishing from $-\infty$ at $T = \varkappa_1$ to zero at $T = \varkappa_1/N$. We shall also see that from $T = \varkappa_1/N$ to $T =$ zero, μ^2 augments from zero to one. All this was I believe developed by Sellmeier ten or twelve years ago. Our molecular theory gives no dynamical foundation for the assumption of μ a mixed real and imaginary numeric, which Cauchy has used for explaining metallic reflection; but it by no means follows that some modified molecular theory may not give some dynamical foundation for this assumption, which acquires great importance, and is at all events rendered exceedingly interesting, by the remarkable success of Cauchy's formulas for metallic reflection even if viewed only as merely empirical.

For the present, however, we confine ourselves to assumptions for which we see a definite dynamical foundation, and of which we can, as it were construct a mechanical model, according to the molecular hypothesis we have been considering. We shall therefore restrict ourselves to μ^2 a real positive or negative integer, and try what we can do towards explaining the translucency of thin metallic films, the known phenomena of metallic reflection, and Kerr's discovery regarding the reflection of light from a polished magnetic pole, by supposing that for metals $-\mu^2$ is a real positive integer, μ^2 according to our notation of equations (37) and (40) above. We do not now assume $r=1$ as it is only for transparent substances that any reason for this supposition has been discovered from observation or theory; and we may imagine that the effective rigidity of the ether acting in the interstices between the molecules, should be largely different from the true rigidity of the homogeneous matter constituting the ether. In fact it is clear that if the round massless sheath of our molecule is infinitely rigid the effective rigidity of the ether in the interstices, would be much greater than the true rigidity of continuous ether; But on the other hand if the sheath of each molecule be not rigid, but more or less yielding and quite perfectly elastic, the effective rigidity of the ether in the interstices might be either greater than, or equal to, or less than, the true rigidity of continuous ether.

Now looking to our formulas (42) and (56) above we see that when $-\mu^2$ is positive, the intensity of the reflected ray is equal to that of the incident ray, both for vibrations, (42), in the plane of the two rays, and for vibrations, (56), perpendicular to this plane. Thus reflection at the surface of a medium for which $-\mu^2$ is positive is total. This totality is for all angles of incidence, and therefore the case is far from being analagous to that of total internal reflection in a transparent medium; the totality in this case being essentially confined to incidence exceeding the critical angle $\sin^{-1}(1/\mu)$. The reflection of light when polarized in the plane of incidence, or perpendicular to it at a well polished silver surface involves, as has long been well known, very little loss of light; about 8 or 10 per cent. has been generally supposed to be the amount of the loss.

Sir John Conroy has shown that the loss is really much less than this, when the metal is very pure and the polish of the surface very perfect. Thus he succeeded in getting so good a polish on a double silver film deposited on glass (Proc. Roy. Soc. of London, May 15, 1884), that with light polarized in the plane of incidence, the loss by reflection was only 2.7 per cent, when the angle of incidence was 30°; and was not discoverable by very delicate observations, and seems to have been proved to have been less than a half per cent. at angles of incidence of from 50° to 75°. With the same reflector and light polarized perpendicularly to the plane of incidence, he found no loss of light at incidences of 30°, and losses of from 2.5 to 6 per cent. at incidences of from 40° to 75°. Whether a somewhat thicker film, or still more perfect polish, would annul these losses, or nearly annul them, is a very interesting subject for inquiry, and it is much to be hoped Sir John Conroy will continue his observations. Meantime we may take silver as a body which is certainly not far from fulfilling the totality of reflection given by our supposition of μ^2 positive, with no assumption of conditions causing the extinction of light. At the same time it is obvious that for any other metal than silver extinction of a large per centage of the incident light is an essential and most serious condition of the problem. It is easy to imagine that our molecular hypothesis can be adapted, without any unnatural straining to directly take into account this condition. For the present, however, I must confine myself to the case of no extinction, and to silver as our one illustration.

Looking now to formulas (42) and (56) we see that for vibrations in the plane of the two rays, the reflected ray is retarded in phase relatively to the incident ray, by an amount which reckoned in radianal measure is equal to $2f-\pi$ while for vibrations perpendicular to the plane of the two rays, the phase of the reflected ray is accelerated relatively to that of the incident ray by an amount $2e$. Hence if the incident light be polarized in any plane oblique to the plane of incidence the reflected ray consists of two plane polarized components, of which the one consisting of vibrations in the plane of incidence, is in phase behind the ether by an amount equal to

$$2f-\pi+2e \quad . \quad . \quad . \quad . \quad . \quad . \quad (57),$$

and by the formulas (44) and (56) for f and e we see that

$$2f-\pi+2e = 2\left\{\tan^{-1}\frac{\kappa}{H}-\tan^{-1}\frac{\cos i}{\mu(V^2+\sin^2 i)^{\frac{1}{2}}}\right. \quad . \quad . \quad (58).$$

The retardation of phase of the component consisting of vibrations in the plane of incidence, relatively to that of the component consisting of vibrations perpendicular to the plane of incidence expressed by this formula, vanishes, as it must do, when $i=0$; because for normal incidence there is no distinction between the two polarized components. If we increase i from 0 to 90°, the retardation increases from zero to π which agrees with observation. If we suppose both v and rv, being the χ of (37) &c., to be very large numerics, we have $h = v$ and therefore by (39)

$$\frac{H}{K} = \frac{\sec i}{rv} + \tan i \quad . \quad . \quad . \quad . \quad . \quad . \quad (59)$$

Hence, with the same approximation in the second term of its second member, (58) becomes

$$2f - \pi + 2c = 2\left\{\tan^{-1}\left(\frac{\sec i}{rv} + \tan i\right) - \tan^{-1}\left(\frac{\cos i}{rv}\right)\right\} \cdot (60)$$

Remark now that unless r be very small rv is very large, and therefore the second member of (60) increases very suddenly, from zero when $i=0$ to being very little short of π when i is still quite small and then completes the small difference of growth up to π a i increases to 90°. This is not consistent with observation and therefore we must suppose r very small, small enough to make rv be of moderate dimensions. For example, if we take $(rv)^{-1} =$ 3.65 we find that the second member of (60) increases probably from 0 $\frac{\pi}{2}$, as i is increased from 0 to 75° 48′ and completes the growth up to π as i is increased further up to 90°. This is precisely the case for silver according to Sir John Conroy's observations (Proc. Roy. Soc. of London, May 15th 1884) the value which he finds for the "principal incidence"* in the case of his double silver film being 75°47′.

*"Principal incidence" is the name technically given by Jamin, Quincke, and others to express the angle of incidence at which the difference of phase of the two reflected components is a quarter of the period. Thus if light be polarized at such an azimuth that the two polarized components the reflected ray are of equal intensity, and if the angle

It is probable that the law according to which the relative retardination increases up to $\frac{\pi}{2}$ and again from $\frac{\pi}{2}$ to π as the incidence increases to 75°. 47′ and again from 75°.47′ to 90°, may be found as accurately expressed by our formula (60) with the value 2.935 for $(r\,V)^{-1}$, as it can be determined by observation: but observation is needed to test this supposition. Should the result shew insufficient agreement with the approximate formula (60) it would to adapt (58 to give the requisite agreement with observation by supposing V not so large as to allow $\sin i$ to be neglected in comparison with it in the expression $V^2 - \sin^2 i$ for h_r^2, a supposition which would also give in (58) a perceptible effect to the terms $-\chi$, u, $u+1$, &c., of (39) which are neglected in (60).

For other metals than silver with the different values of the principal incidence found for them by observation, it would be also easy by the approximate formula (60) to find values of rV which would give the observed value for the principal incidence and if necessary to introduce the necessary modification by the more complete formula (58) to obtain agreement with observation. It is interesting to observe that the general law of metallic reflection, which has been found by observation, according to which the component of the reflected light whose plane of polariza- is perpendicular to the plane of incidence is retarded relatively to the other component by an amount which augments from zero to π as the angle of incidence increases from zero to 90°, is brought out without any strained

---

incidence be the principal incidence, the reflected light is circularily polarized. The azimuth thus defined for light at the principal incidence is called the "principal azimuth." The reflection being so nearly total as we have seen it to be the principal azimuth for the silver surface ought to be very nearly 45°. Sir John Conroy's measurement of principal azimuth gives 44° for his double silver films.

supposition by our formulas (58) provided the direction of vibration be perpendicular to the plane of polarization as we have been compelled by other reasons to believe it to be.

It seems then as if we might be very happy in our molecular explanation of metallic reflection; but alas, one most serious and seemingly essential characteristic of metallic reflection remains unexplained and that is the fact that there is in it very little of what we might call chromatic dispersion; which in this case would shew itself in differences of the principal incidence for light of different periods. Our dynamical theory makes $\nu^2+\nu$ vary for different colors approximately in proportion to T_1^2 when T is very small compared with κ_1. We have no dynamical theory advanced enough to give the law of relation between $\mathfrak{n}_1$ (the effective rigidity of the ether in the interstices between the molecules) and the period of the vibrations. It is difficult to conceive how any natural or acceptible theory could bear the strain of being forced to make the produce $\rho\nu$ as nearly the same through the wide range of period presented by the different colours of visible light as is necessary to account for the known facts of metallic reflection. We are thus forced to admit that our dynamical theory of metallic reflection is a failure for the present but it is not unsuggestive and it may possibly help to the true dynamical explanation which is so much desired. That it does indeed contain part of the essence of the true dynamical theory, can scarcely be doubted after we have considered the next two subjects on which we are going to try it: the translucency of thin metallic films, and the effect of magnetism on polarized light incident on polished magnetic poles, or transversing thin films of magnetized iron, nickel or cobalt. The three remarkable discoveries of Quincke, Kerr and Kundt in this subject are as we shall see all brought out directly and without strain from our molecular theory.

Translucency of Thin Metallic Films.*

To avoid circumlocution we shall continue to use the words "upper" and "lower," and suppose the light to be incident in our upper medium, with a horizontal interface between it and a denser medium below the interface. We shall now suppose the denser medium to be in the form of a plate between two parallel faces, and the medium below the plate to be the same as the medium above it. There is no difficulty in working out by the general method expressed in the equations (1), (53) and (54) above, the problem for the reflection of light from the plate into the upper medium, and the transmission of light through the plate into the lower medium, for the two cases of vibrations perpendicular to the plane of the three rays, and vibrations in this plane. If we work this out for either case and for μ real, we find with great ease the ordinary formula expressing the wave theory of Newton's colors of thin plates. The only difference between the two cases is, that the intensity of the reflected light for a simple reflection at one surface, varies differently with the angle of incidence in the two cases. The complication of different acceleration or retardation of phase at different incidences, presented by the case of vibration in the plane of the three rays, does not involve any additional complication, when we pass from reflection and refraction at a single interface, to the problem of the plate.

Working out the problem for $-\mu^2$ real and positive, and equal to ν^2 as above, and taking

δ to denote the thickness of the plate,

$x = 0$ to correspond to its upper side,

and $x = -\delta$ to correspond to its lower side,

we find as follows for the whole motion of the mediums due to a plane wave incident in the upper medium; all our

* Added. Dec. 4–11, 1884.

other notations being the same as before; and now for brevity put

$$q = by + \omega t \qquad (61)$$

Upper Medium $\left\{ \xi = \frac{1}{2} \sec e \left\{ \underbrace{\cos(ax+q+e) - \varepsilon^{-2k\delta}\cos(ax+q+3e)}_{\text{Incidient Wave}} + \underbrace{(1-\varepsilon^{-2k\delta})\cos(-ax+q+e)}_{\text{Reflected wave}} \right\} \right.$

Motion in plate, $\xi = \varepsilon^{kx}\cos q - \varepsilon^{-k(2\delta+x)}\cos(q+2e)$

Wave transmitted into lower medium $\Big\} \; \xi = 2\sin e . \varepsilon^{-k\delta} \cos\left[a(x+\delta)+q+e-\frac{\pi}{2}\right]$ $\Bigg\} (62)$,

where as in (37) and (56) above

$$k = \sqrt{(v^2 + \sin^2 i)} \qquad (63)$$

and

$$\tan e = \frac{2k}{a} \qquad (64).$$

Taking only this last equation into account, it is easy to verify that equations (62) fulfil, at each interface the proper interfacial conditions which are that on the two sides of each interface the values of ξ are equal, and the values of $n\frac{d\xi}{dx}$ in the plate equals the value of $\frac{d\xi}{dx}$ in the contiguous medium on the other side of the interface.

<u>Reflection from and transmission through a plate for the case of vibrations in the plane of the three rays</u>

The result so far as the waves in the upper and lower mediums must clearly be identical with that expressed in (62) with $\frac{\pi}{2} - f$ substituted for e; f, as in equations (41) and (39) above being found by the following formula

$$\tan f = \frac{-\chi + (\chi+u)^2 - \frac{k}{f}u^2}{\frac{k}{a}\left[\chi - (1+u)^2\right] + \frac{b}{a}(\chi+u+1)^2} \qquad (65).$$

The corresponding value of the four coefficients corresponding to B and B' of (38), which are now required to express the double interfacial wave are easily written down by aid of (38) but they are not required for our present purpose.

Looking back now to (62), whether with e as in the formula for vibrations perpendicular to the plane of the three rays, or with $\frac{\pi}{2} - f$ in place of e to suit the case of vibrations in the plane of the three rays, we see that when

δ is infinitely small the reflected wave vanishes, and the wave transmitted into the lower medium agrees with the incident wave in amplitude and phase: that is to say the film has no effect which is of course the correct result for this case.

Next suppose δ to be large enough to make $\varepsilon^{-k\delta}$ exceedingly small. The wave transmitted into the lower medium becomes infinitely small and the reflected wave in the upper medium agrees infinitely nearly with what we found above in (42) and (56) for the case of reflection at a single metallic surface.

When $\varepsilon^{-k\delta}$ is a small fraction of unity, not zero, the ~~amplitude of the transmitted wave~~ the amplitude of the transmitted wave is approximately $4 \sin e \cos e \, \varepsilon^{-k\delta}$ of the amplitude of the incident wave for the case of vibrations perpendicular to the plane of the three rays; and the same with f for e for the case of vibrations in this plane. The phase of the transmitted wave is accelerated by an amount approximately equal to $a\delta + 2e - \frac{\pi}{2}$ in the former case, and equal to $\frac{\pi}{2} - f$ in the latter case. The amount of the acceleration thus calculated for each case is that by which the transmitted wave is in advance of an ideal continuation of the incident wave with the plate removed. The unit of reckoning is the radian. To reduce to space travelled in the medium on either side of the plate we must divide by $a \sec i$. Hence remarking that

$$a = \frac{2\pi}{\lambda} \cos i \qquad (66);$$

where λ is the wave ~~in the~~ length in the medium on either side of the plate, we find for the amounts of the advance of phase in the two cases:—

vibrations perpendicular to the plane of the three rays $\Big\}$ $\cos i . \delta + \left(\frac{e}{\pi} - \frac{1}{4}\right)\lambda$ . . . . . (67);

vibrations in the plane of the three rays $\Big\}$ $\cos i . \delta + \left(\frac{1}{4} - \frac{f}{\pi}\right)\lambda$ . . . . . (68).

We have seen that when $i = 0$, e and f are each positive acute angles and complements of one another; and each augments to the value $\frac{\pi}{2}$ when i is increased from 0° to 90°. Hence the second members of (67) and (68) vanish respectively for two particular values of i. In these cases the advance of the corresponding polarized component is equal to cos i.

§. To explain this let ab be a wave front in the upper medium and $a'b'$ the position it will reach in the lower medium after any particular time t. Now imagine the plate to be annulled and the lower medium to be moved perpendicularily to the plane of the plate so as to fill up the gap. The phase of the transmitted wave $a'b'$ in its actual position is the same as would be the phase at the same time t, in the altered position of $a'b'$, with the plate annulled.

When the second term of (67) or (68) is positive, there is an advance of the transmitted ray even more than that corresponding to the annulment of the plate. There is positive advance, though of less amount than corresponding to annulment of the plate, when the second term is negative, but of less absolute value than the first. The general result of advance of phase produced by a metallic film upon light transmitted through it was discovered experimentally by Quincke 21 years ago; but alas for our dynamics, the details of his results seem very far from agreeing with anything I can make out of our formulas. We must not however be discouraged by this. At all events the nearest approach to the explanation of Quincke's result, on the supposition of a real refractive index, makes the refractive index vary with the angle of incidence — a brilliant reductio ad absurdam; and gives it values ranging from .3 to 8 or 9 for different metals or even for different specimens of the same metal!

Our dynamical theory perfectly explains Kerr's result, for normal reflection from a metallic pole, crossed, whether normally or obliquely, by lines of magnetic force; which is, that plane polarized light incident normally or nearly normally, produces a plane polarized reflected ray, with plane of polarization turned slightly in the direction opposite to that of the "Amperian currents" of the magnetisation. The effect of magnetisation of the iron must be to give different values to v for circularily polarized light, according as the direction of the orbital motion is with or against the "Amperian currents." Thus while, according to our formulas, there is for every ray total reflection the effect of the magnetisation is to change, in the act of reflection not the intensity but the phase of circularily polarized ray. Hence plane polarized light incident nor-normally, ideally resolved into two opposite circularily polarized rays, gives rise to two opposite circularily polarized reflected rays, differing slightly in phase and therefore equivalent to a plane polarized ray in a plane of polarization turned through a small angle. On the other hand, if we imagine the iron to act as a transparent medium, with real refractive index, the only possible effect in the case of normal incidence is to give different intensities to the two circularily polarized components of the reflected ray, and so to give a slight degree of ellipticity to the reflected ray, with major axis of the ellipse precisely coincident with the line of vibration of the incident light. This is Fitzgerald's result, which, as remarked by Fitzgerald himself, and by Kundt, is absolutely at variance with Kerr's experimental discovery. It is therefore quite certain that iron does not act as a transparent medium with real refractive index. It is, however, quite conceivable that the extinctivity which the iron must have (to give it its practical opacity), if it has a real refractive index, may, under the influence of magnetisation, give the difference of phase required to explain Kerr's result. That extinctivity must indeed be invoked (as Cauchy long ago invoked it) seems in this new case probable, because though our dynamical formulas, without extinction perfectly explain Kerr's result they are utterly at variance with Kundt's,* according to which the plane of polarization of light passing normally through a thin iron plate, is turned through a not very small angle (amounting in some of his experiments to as much as $3\frac{3}{4}°$ for iron, 2° for cobalt and $\frac{3}{4}$ for nickel), in the *opposite* direction to that found by Kerr in reflection. But it does not seem possible to abandon our pure imaginary for refractive index in metals ($-\mu^2$ real), and we may hope that extinctivity on a true dynamical foundation in connection with our molecular theory, which it must be remembered is due originally to Sellmeier, may serve to solve the numerous difficulties in connection with metallic reflection and transmission, which give us so much anxiety. Extinctivity however cannot help to solve the great difficulty as to reflection at the interface between two transparent mediums, in the case of vibrations in the plane of the three rays. Green's attempt to explain this difficulty by gradualness in the transition of physical quality from one medium to another seems to me most unpromising if not utterly hopeless. There remains Green's other suggestion of "extraneous force," by which as we have seen he opened a door for explaining how the velocity of light in a crystal can depend on the direction of the line of vibration, irrespectively of the line of propagation. If this suggestion becomes realized it must modify the circumstances at the interface which determine the reflection. Is it possible that it can lead to the true law for reflection of waves consisting of vibrations in the plane of the three rays?

---

* Berlin Sitzungsberichte July 10, 1884 or Philosophical Magazine October, 1884

In continuation of MSS "Improved Gyrostatic Molecule" despatched 1st Nov. 1884.*

and the number greater and greater, with the same total mass, the angular velocity γ must be augmented in inverse proportion to k^2.

Now for our improved gyrostatic molecule imagine two kinetically equal and similar rotators mounted by means of ball and socket joints in the interior of a rigid spherical sheath; and, by axles projecting from them towards the centre of the sheath let them be jointed together in the manner indicated; that is to say by a ball projecting from the one fitting in a cylindrical projection from the other. To make them kinetically equal and similar as supposed, notwithstanding this slight difference of form their masses and moments of inertia round corresponding axes are to be exactly equal. To avoid all complexity we shall suppose the outsides of the sheath to be perfectly smooth and of truly spherical figure, so that when embedded in ether it may not be affected by the rotational part of the motion of the ether, and that it may experience merely translational forces in lines through its centre in virtue of the translational motion of the ether. We shall however in investigating the kinetic properties of our new compound molecule not restrict ourselves to the supposition of perfect smoothness in the sheath, and shall consider the result of the giving of any motion, whether translational or rotational to the sheath.

First suppose the interior rotators to be given at rest with their axes in one line as indicated by the strong lines in the diagram.

The diameter of the molecule through the centre of the ball and socket joints will, for brevity be called the

*Added Dec 11 to 13, 1884, continued from page 293.

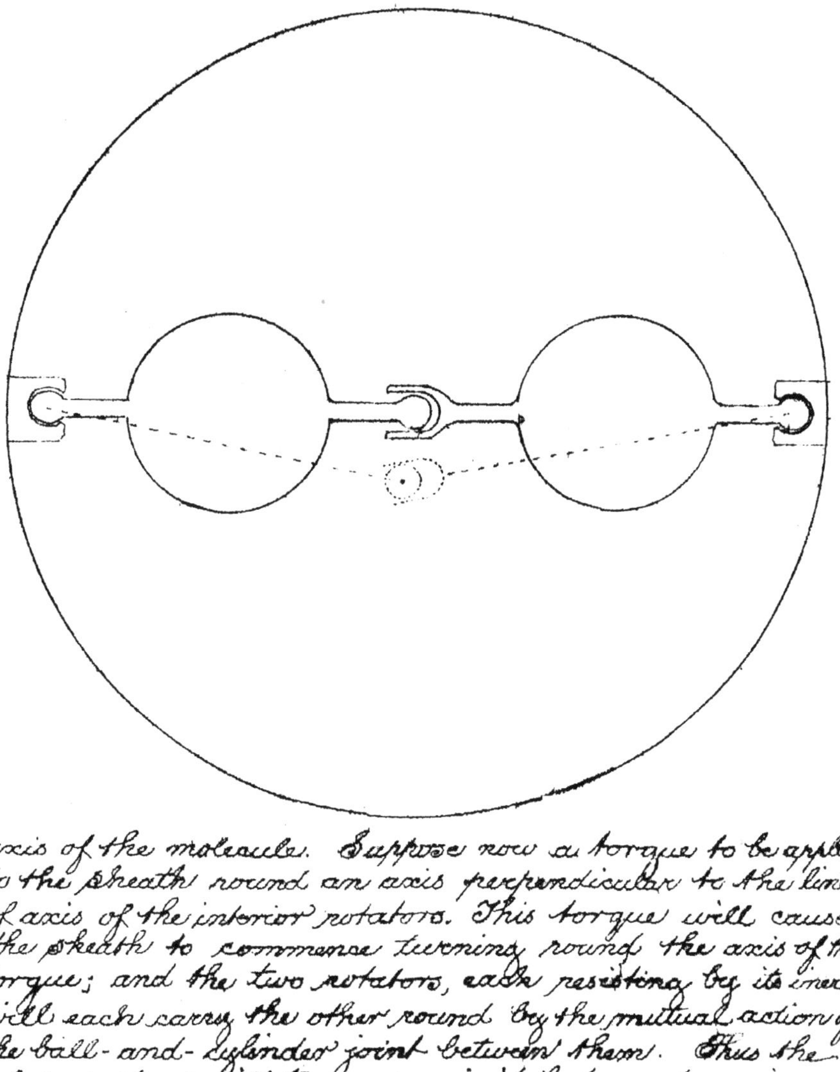

axis of the molecule. Suppose now a torque to be applied to the sheath round an axis perpendicular to the line of axis of the interior rotators. This torque will cause the sheath to commence turning round the axis of the torque; and the two rotators, each resisting by its inertia, will each carry the other round by the mutual action of the ball-and-cylinder joint between them. Thus the whole system will turn as a rigid body, and receive acceleration from the supposed torque according to the law of acceleration of a rigid body. Suppose now that a force be applied to the sheath in a direction perpendicular to the axes of the rotators. They will clearly lag in the motion thus produced, and their axes turning in opposite directions will make an increasing obtuse angle with one another till the rotators strike the sheath. It is curious to see how this mode of jointing gives perfect quasi-rigidity relatively to rotatory motion of the sheath and absolute limpness to all translational motion of the sheath except along the line of axes of the rotators (which, be it remembered, we had initially in one line). Suppose now the two rotators, given with their axes in one line, to be set into rapid rotation round this line. The quasi-rigidity relatively to rotation of the sheath still remains perfect; and therefore for all rotational motions of the sheath with its centre unmoved, the rotators will act precisely as if they were rigidly connected; so that the compound molecule will act merely as a simple gyrostat.

Relapsing now onto the supposition of the sheath perfectly smooth on its outside let any forces be applied to it. It is clear that its motion will be purely translational when we consider the symmetry of the reactions in the two ball-and-socket joints. A result of any acceleration of the centre of the sheath not exactly along the axis of the molecule must be to disalign the axes of the rotators; but if the regular velocity of the rotators be very great, their gyrostatic action will give rise to an exceedingly great quasi-rigidity against the disalignment. It is easy to write down the equations of the translational motion of the sheath and of the whole motion rotational and translational of the rotators, under the influence of any given forces, applied normally as supposed to the sheath. For our present purpose it will be sufficient to write down these equations of motion for the case of infinitesimal disalignment of the axes of the rotators, but it will help us to understand all the circumstances if we first take the rigorous solution for the case of steady precessional motion of the rotators with their axes inclined at any finite angle θ, to the axis of the molecule. This steady motion involves uniform circular motion of the sheath; or in one particular case zero motion of the sheath.

OI is perpendicular to OB

BOA = θ ; BOI = u;

γ = component angular velocity round OB;

ζ = component angular velocity round OI;

ω = angular velocity of the plane BOI round OA.

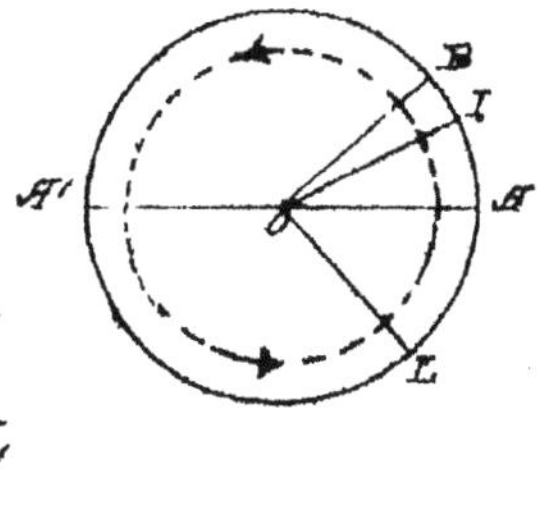

Let O be the centre of one of the vibrators, and A'OA a line through it parallel to the axes of the molecule. This, on account of the symmetry, is the line joining the centres of the two rotators. Let OB and OI be respectively the axis of figure of the rotator and its instantaneous axis of revolution at any time. The supposed motion will be the same as that of a cone having OB for its axis and BOI for its semi-vertical angle rolling on a fixed cone having OA for its axis and IOA for its semi-vertical angle (compare Thomson & Tait's Natural Philosophy § 105). Suppose now the component angular velocity round OB to be of any given magnitude γ. This remains absolutely constant because the ball-and-socket and ball-and-cylinder joints are perfectly frictionless. Suppose now in the case of motion investigated the angle BOA to be given equal to θ and the precessional angular velocity of given magnitude ω. Draw OL perpendicular to OB in the plane BOA, and let it be required to find the component angular velocity of the rotation round OL which we shall denote by ζ

Make OL, OA, OI, OB each equal to unity. The linear velocities of the matter at I and at B, are respectively equal to the angular velocities γ and ξ. Now the same matter is also at the point B, and therefore the required angular velocity ξ is simply equal to the linear velocity of the point B in the diagram. Supposing the rotational and precessional motions viewed from A to lie in the direction opposite to the hands of a watch, B and I move perpendicularily to the paper outwards, and I perpendicularily to the paper inwards as follows (the second expression for the velocity of B being found by considering that the velocity of B is also the velocity of the matter of the rotator at B, and that the velocity of the matter of the rotator at I is zero):–

$$\xi = \text{linear velocity of } B = \omega \sin\theta = \sqrt{(\gamma^2 + \xi^2)} \sin u \quad \ldots \quad (13)$$
$$= \gamma \tan u$$

$$\text{linear velocity of } I = \omega . \sin(\theta - u \quad \ldots \quad (14)$$

$$\text{linear velocity of } L = \omega . \cos\theta \quad \ldots \quad (15)$$

Of these expressions the only ones we require are the first, for ξ, and the third, for the velocity of L. The others are put down merely to illustrate the circumstances.

Let m be the mass of the rotator and mk^2 and ml^2 its moments of inertia respectively round OB and OL. The component moments of momentum round the axes OB and OL are respectively $mk^2\gamma$ and $ml^2\xi$. Hence as the points B and L have absolute velocities perpendicular to the plane of the paper outwards and inwards equal respectively to ξ and $\omega \cos\theta$, the moments of the couples required to produce the corresponding changes of direction of the two components of momentum are respectively $mk^2\gamma\xi$ and $m.l^2\xi.\omega.\cos\theta$. These couples are both in the plane of the diagram, the first in the direction of the arrow heads and the second in the opposite direction and therefore the whole couple required to cause the rotator to move as it does is

$$m(k^2\gamma - l^2\omega\cos\theta)\xi = m(k^2\gamma - l^2\omega\cos\theta)\,\omega\sin\theta \quad \ldots \quad (16)$$

Now let us suppose the sheath of the compound molecule to be kept so moving that the centres of the ball-and-socket joints revolve with ~~with~~ uniform angular velocity ω, in circles each of radius r, perpendicular to the axis of the molecule; and let it be required to find what must be the value of θ, in order that the rotator may move with steady preressional motion in the manner supposed. Let F denote the force towards the centre of each of these circles, with which the socket acts upon the ball turning within it. Imagine after Poinsot pairs of equal balancing forces F, to be applied in parallel lines through the centres of inertia of the two rotators, we thus have a force F at the centre of inertia of each rotator, and a couple whose moment is $F a \cos\theta$, if $\underline{a}$ denote the distance from the centre of the ball-and socket joints to the centre of inertia of the rotator. The centre of inertia of each rotator in these circumstances, moves with angular velocity ω, in a circle whose radius is $r + a\sin\theta$; and the centreward force required to cause it to so move (or the force balancing its centrifugal force) is F. Hence

$$F = m\,\omega^2 (r + a\sin\theta) \quad . \quad . \quad . \quad (17).$$

The function performed by the couple is to change directions of moments of momentum in the manner explained above, and therefore it must be equal to the formula (16):–

$$F a \cos\theta = m(k^2\gamma - l^2\omega\cos\theta)\,\omega\sin\theta \quad . \quad . \quad (18)$$

These two equations serve to determine F and θ.

For our present purpose it is sufficient to work out the result for θ infinitely small. Thus by taking θ, for $\sin\theta$, and 1 for $\cos\theta$, in (17) and (18), we find

$$\theta = \frac{r a \omega}{k^2\gamma - (a^2 + l^2)\omega} \quad . \quad . \quad . \quad . \quad (19);$$

and

$$F = m\,\omega^2 r \left[1 + \frac{a^2\omega}{k^2\gamma - (a^2 + l^2)\omega}\right] \quad . \quad . \quad . \quad (20)$$

We conclude that for circularly polarized light the effect of rotation within the gyrostatic molecule is to cause it to have the same influence on the motions of the ether; as if its mass instead of $2m$, were

$$2m_{,} = 2m\left[1+\frac{a^2\omega}{k^2\gamma-(a^2+l^2)\omega}\right] \quad . \quad . \quad . \quad (21)$$

Hence supposing $\rho+m_{,}$ and $\rho+m'_{,}$ to be the effective density of the ether with its embedded molecules, for two circularly polarized rays with opposite orbital motions, and $v\,v'$ the velocities of propagation of these rays, we have

$$\frac{v}{v'} = \sqrt{\frac{\rho+m_{,}}{\rho+m'_{,}}} \quad . \quad . \quad . \quad . \quad . \quad (22);$$

where $m'_{,}$ is the same as $m_{,}$ as given by (21), with the sign of ω changed. Hence, the ratio being exceedingly nearly equal to unity, we have approximately

$$\frac{v}{v'} = 1+\frac{1}{2}\left[\frac{a^2\omega}{k^2\gamma-(a^2+l^2)\omega}+\frac{a^2\omega}{k^2\gamma+(a^2+l^2)\omega}\right] \quad . \quad . \quad (23).$$

If ω could be large enough to make $(a^2+l^2)\omega$ equal to or greater than $k^2\gamma$, we should have something analagous to "anomalous dispersion" in the magneto-optic effect. It does not however appear probable that any such critical condition can be at all approximated to by the highest ultra-violet light known to exist; and for the present it is convenient to suppose $(a^2+l^2)\omega$ infinitely small in comparison with $k^2\gamma$, which reduces (23) to

$$\frac{v}{v'} = 1+\frac{a^2\omega}{k^2\gamma} \quad . \quad . \quad . \quad . \quad . \quad . \quad (24)$$

This as worked out in (8) to (12) above, leads to the true law of relation between the rate of turning of the plane of polarization, and the period of vibration of the light. It is very curious to remark that the gyrostatic efficiency of our improved double-rotator molecule, depending as it does on translational, not on rotational, motion of the sheath, is inversely proportional to the angular velocity of the rotators; provided this angular velocity be great enough for gyrostatic domination (Thomson & Tait's Natural Philosophy, second edition, § 345): while the gyrostatic efficiency of our crude original gyrostatic molecule, (depending as it does on the rotational motion of the sheath) was directly proportional to the angular velocity of the rotator.

Going more into detail we see that with the crude original gyrostatic molecule, the proportional alteration of the velocity of light due to circular polarization depends on $\frac{\omega \gamma k^2}{v^2}$: whereas, with the improved gyrostatic molecule, it depends simply on $\frac{\omega}{\gamma}$ (really upon $\frac{\omega a^2}{\gamma k^2}$, but we may suppose $\frac{a^2}{k^2}$ to be some constant numeric of moderate value a little more or less than unity). If now the improved gyrostatic molecule, instead of being perfectly smooth on the outer boundary of its sheath, as for simplicity we took it in the investigation, be now supposed to be adhesively embedded in the ether, so that it shall be carried ~~round~~ with the ether in the infinitesimal rotations which the ether experiences in the course of luminiferous vibrations, it will act in respect to these rotations as if it were a simple vibrator like the unimproved molecule, and at the same time it will have efficiency in virtue of its translatory motion, according to the result of the preceding investigation — the same efficiency in respect to translational movements as if its outer surface were smooth as supposed in the investigation. And now what is most important we see that if the linear dimensions of the molecule be made small enough, without changing the angular velocity of its ~~rotators~~, the influence of the rotational motion on the sheath becomes smaller and smaller, and quite insensible in comparison with the gyrostatic effect due to translational motion of the sheath; this last remaining unchanged with the diminution of linear dimensions, provided that not only the angular velocity, but the ratio of the mass of the rotators to the whole mass of the molecule is kept

$z^2 = \frac{1}{\eta^2}$	$-\frac{\xi}{x} = y$	$z = \frac{1}{\eta}$
SEVENTH BRANCH.		
.0	−1.7678	.0
.001	−1.3944	.03162
.00125	−1.1781	.03536
.0014	−0.7273	.03741
.0014701	+0.000023	
.0015	+0.8512	.03873
SIXTH BRANCH.		
.0016	−5.41	.04000
.0018	−2.235	.04242
.002	−1.9564	.04472
.0025	−1.7665	.05000
.005	−1.5128	.07081
.007	−0.7104	.08366
.0072	−0.2302	.08485
.0072564	+0.000070	
.0075	+3.4476	.08660
FIFTH BRANCH		
.008	−3.82	.08944
.009	−2.13	.09487
.01	−1.922	.1
.015	−1.6676	.12247
.016	−1.5956	.1264
.02	−1.4001	.1414
.022	−1.2172	.1483
.025	−0.4366	.1581
.0255607	+0.0000621	
.0265	+1.9560	.1628
FOURTH BRANCH		
	−12.268	.1700
	−2.731	.1750
.03̇	−2.278	.1826
.04	−1.9235	.2
.05	−1.7130	.2236
.07	−1.315	.2640
	−0.983	.28
.08801	+0.00042	
.09	+0.4997	.30
	+2.086	.305

$z^2 = \frac{1}{\eta^2}$	$-\frac{\xi}{x} = y$	$z = \frac{1}{\eta}$
THIRD BRANCH.		
.1	−9.3911	.3162
	−5.66	.32
.1122	−3.0162	.335
.1225	−2.454	.35
.16	−1.877	.40
.25	−1.0385	.50
.28	−0.640	.5292
.29	−0.278	.5385
.29849	+0.000.06	
.3136	+0.7714	.56
.3̇	+4.90	.5773
SECOND BRANCH.		
.36	−20.38	.60
.4225	−3.450	.65
.49	−2.460	.7
.5625	−2.001	.75
.7225	−1.345	.85
1.00483	−0.000004	
1.21	+2.643	1.1
FIRST BRANCH.		
1.44	−22.13	1.2
1.69	−4.38	1.3
1.96	−2.66	1.4
2.56	−1.244	1.6
2.89	−0.726	1.7
3.4618	−0.000045	1.8605
3.5	+0.0463	1.8708
4.	+0.631	2.
5.	+1.731	
10.	+6.89	
20.	+16.95	

For rest of branch the equation becomes $y = z^2 - \frac{1}{z^2} - 3$, and the curve is nearly a parabola.

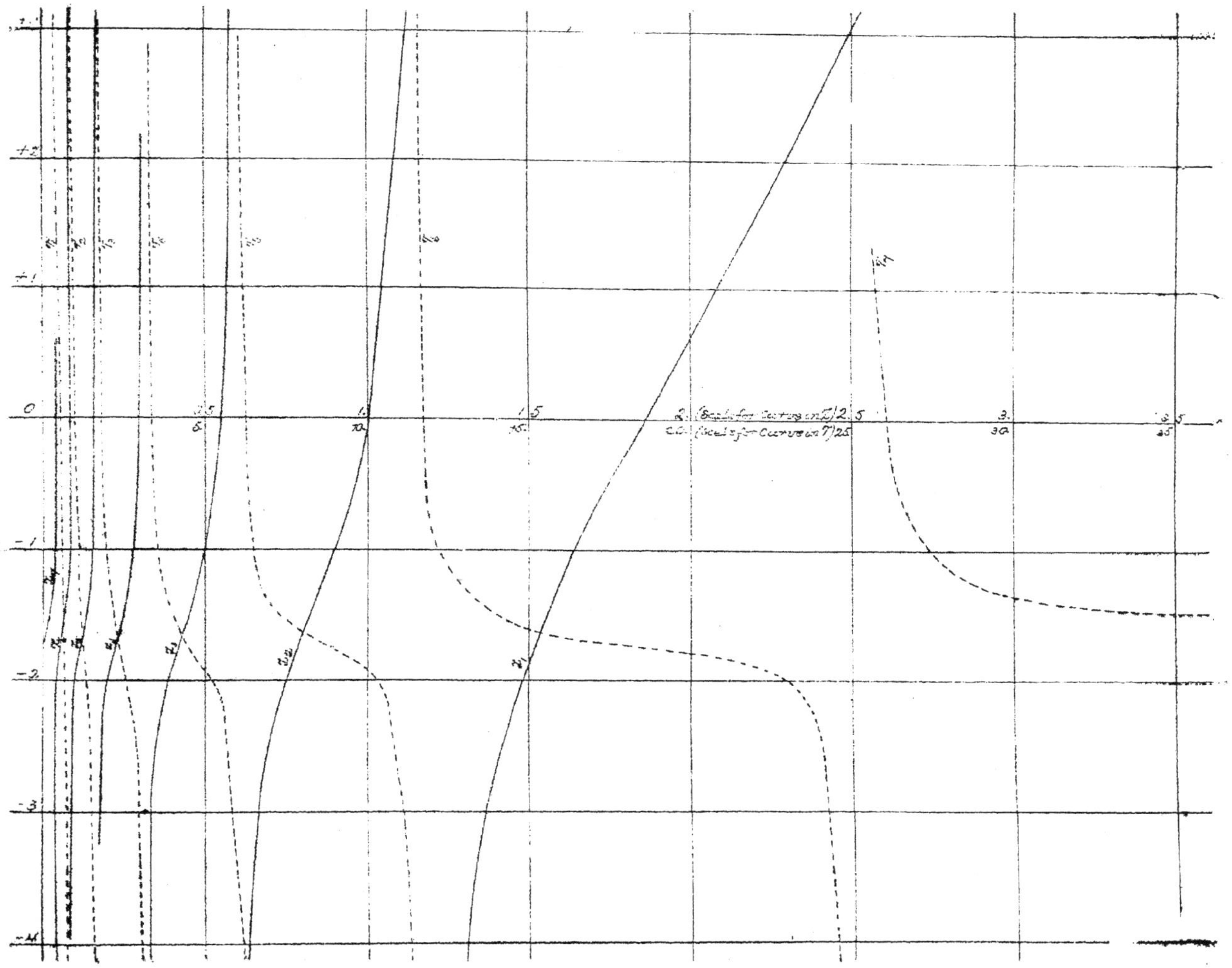

2. (Scale for Curves in Z) 2.5
20. (Scale for Curves in η) 25

unchanged.

The kinetic properties of the improved gyrostatic molecule are exceedingly interesting, but we have had all of them that are essential to our present purpose and time presses so much that I close this final despatch without even writing down the cartesian equations of its motion.

Since my return, Prof. E. W. Morley has kindly sent me a diagram of curves giving a complete graphical representation of $-\frac{\xi}{x}$, [for the problem proposed on page 103 of the Lectures] both as a function of T and $\frac{1}{T} = z$. This I hope to make good use of in attempting to explain Extinction, Anomalous dispersion, and Fluorescence and Phosphorescence. The results of Prof. Morley's calculations are here appended. W.T.

[The table of roots of $\frac{-\xi}{x} = 0$ and their ~~corre~~sponding displacement and energy ratios is given on page 251. Another table is here added for branches of the curve in z. The branches are numbered so as to bring the corresponding fundamental periods, $\varkappa_1, \varkappa_2 \ldots \varkappa_7$, in ascending order of magnitude. These branches are given by full lines in the diagram. The dotted lines are for the reciprocal branches in T, which are drawn upon a longitudinal scale $\frac{1}{10}$ of that in the former set of curves so as to bring the two sets upon the same diagram. H.]

INDEX OF LECTURES.

(With some references in connection with the subject matter)

LECTURE I. 1

Kinetic Theory of Gases.

Boltzmann, Weiner Akad. Vols. 58, 89, 90. Clausius Pogg. Annal. #10, 1858, Phil. Mag. 1859, 1860 Maxwell Phil Mag. 1860, Camb. Phil. Trans. XII, 1879, Ency. Brit. 9 Ed. "Atom"

Molecular Dynamics [and Vortex Atoms].

Ball, Phil. Mag. Vol. 36, 1868, Berthat C. R. Vol. 98. Cauchy, Mem. Acad. Sci. Paris 1850, C R. 1847–49, 50 Craig, Am. Jour. Maths 1880. Glazebrook Phil. Mag. June 1881. Helmholtz, Wirbel bewegung "Wiss Abh" Crells Vol. 55, 1885, Phil. Mag. 38, 1867. Lagrange "Mecanique Analitique", part 2, p. 339. Lamb's "Motion of Fluids". Maxwell's Elec. Mag. Art. 822, Ency. Brit. "Atom". Poisson, Ann. de Chimie 1828–9, Mem. Acad. Sci. 1829. Rankine, Phil. Mag. 39, 1870, Trans. R. S. Edinb. Vol. 25. Stokes, Math. & Phys. Papers, I, II, III. Tait, Molecular Forces, Proc. R. S. Edinb. XII, Trans. R. S. Edinb. XXV, XXVIII, Nat. XXX. Tyndall, Phil. Trans. & Mag. 1864. Thomson J. J. Motion of Vortex Rings, MacMillan & Co. 1883. Thomson Sir Wm Math. & Phys. Papers. Phil. Mag. 1867, 1880. Glas. Phil. Soc. Proc. 1868, Edinb. R. S. Proc. 1869, 1872, 1875-6. Edinb. R. S. Trans XXV, 1869, Brit. Assoc. 1880, 1884. Nat. 1880, 1884, "On the Size of Atoms" Am. Jour. Sci. 1870. Thompson & Tait Natural Philosophy.

The Wave Theory of Light.

Beer, "Höhre Optik". Cauchy, "Oeuvres completes", Paris, Mem Acad. Sci. 1830, '31. C. R. 1836, '38, '39. Fresnel "Oeuvres Completes" Annal de Chimie 1816–46, Pogg. Ann. 1825–48. Green, Math. Papers, Camb. & Edinb. Trans. 1828–41. Glazebrook, "Physical Optics", Appleton & Co. Lorenz, "Lehre vom Licht". Rankine Mis Sci. Papers

LECTURE III. 31

LECTURE IV.

LECTURE V

LECTURE VI.

LECTURE VII. . . . 68

LECTURE VIII.

LECTURE IX.

LECTURE X.

LECTURE XI.

LECTURE XII.

LECTURE XIII.

LECTURE XIV.

LECTURE XV.

LECTURE XVI. 194

LECTURE XVII.

LECTURE XVIII

LECTURE XIX.

LECTURE XX.

APPENDIX.

Printed in Dunstable, United Kingdom

91220006R00194